Lecture Notes in Computer Scien

T0238015

Commenced Publication in 1973
Founding and Former Series Editors:
Gerhard Goos, Juris Hartmanis, and Jan van Leeuwen

Selim G. Akl   Cristian S. Calude
Michael J. Dinneen   Grzegorz Rozenberg
H. Todd Wareham (Eds.)

Unconventional Computation

6th International Conference, UC 2007
Kingston, Canada, August 13-17, 2007
Proceedings

 Springer

Volume Editors

Selim G. Akl
Queen's University
School of Computing, Kingston, Ontario K7L 3N6, Canada
E-mail: akl@cs.queensu.ca

Cristian S. Calude
Michael J. Dinneen
University of Auckland
Department of Computer Science, Auckland, New Zealand
E-mail: {cristian, mjd}@cs.auckland.ac.nz

Grzegorz Rozenberg
University of Colorado
Department of Computer Science, Boulder, Co 80309-0347, USA
E-mail: rozenber@liacs.nl

H. Todd Wareham
Memorial University of Newfoundland
Department of Computer Science, St. John's, NL, Canada
E-mail: harold@cs.mun.ca

Library of Congress Control Number: 2007932233

CR Subject Classification (1998): F.1, F.2

LNCS Sublibrary: SL 1 – Theoretical Computer Science and General Issues

ISSN   0302-9743
ISBN-10   3-540-73553-4 Springer Berlin Heidelberg New York
ISBN-13   978-3-540-73553-3 Springer Berlin Heidelberg New York

Springer is a part of Springer Science+Business Media

springer.com

© Springer-Verlag Berlin Heidelberg 2007
Printed in Germany

Typesetting: Camera-ready by author, data conversion by Scientific Publishing Services, Chennai, India
Printed on acid-free paper   SPIN: 12088782   06/3180   5 4 3 2 1 0

Preface

The Sixth International Conference on Unconventional Computation, UC 2007, organized under the auspices of the EATCS by the Centre for Discrete Mathematics and Theoretical Computer Science (Auckland, New Zealand) and the School of Computing, Queen's University (Kingston, Ontario, Canada) was held in Kingston during August 13–17, 2007. By coming to Kingston, the International Conference on Unconventional Computation made its debut in the Americas.

The venue for the conference was the Four Points Hotel in downtown Kingston on the shores of Lake Ontario. Kingston was founded in 1673 where Lake Ontario runs into the St. Lawrence River, and served as Canada's first capital. Renowned as the fresh-water capital of North America, Kingston is a major port to cruise the famous Thousand Islands. The 'Limestone City' has developed a thriving artistic and entertainment life and hosts several festivals each year. Other points of interest include Fort Henry, a 19th century British military fortress, as well as 17 museums that showcase everything from woodworking to military and technological advances.

The International Conference on Unconventional Computation (UC) series, https://www.cs.auckland.ac.nz/CDMTCS/conferences/uc/, is devoted to all aspects of unconventional computation, theory as well as experiments and applications. Typical, but not exclusive, topics are: natural computing including quantum, cellular, molecular, neural and evolutionary computing; chaos and dynamical system-based computing; and various proposals for computations that go beyond the Turing model.

The first venue of the Unconventional Computation Conference (formerly called Unconventional Models of Computation) was Auckland, New Zealand in 1998; subsequent sites of the conference were Brussels, Belgium in 2000, Kobe, Japan in 2002, Seville, Spain in 2005, and York, UK in 2006.

The titles of volumes of previous UC conferences are the following:

1. C. S. Calude, J. Casti, and M. J. Dinneen (eds.). *Unconventional Models of Computation*, Springer-Verlag, Singapore, 1998.
2. I. Antoniou, C. S. Calude, and M. J. Dinneen (eds.). *Unconventional Models of Computation, UMC'2K: Proceedings of the Second International Conference*, Springer-Verlag, London, 2001.
3. C. S. Calude, M. J. Dinneen, and F. Peper (eds.). *Unconventional Models of Computation: Proceedings of the Third International Conference, UMC 2002*, Lecture Notes in Computer Science no. 2509, Springer-Verlag, Heidelberg, 2002.
4. C. S. Calude, M. J. Dinneen, M. J. Pérez-Jiménez, Gh. Păun, and G. Rozenberg (eds.). *Unconventional Computation: Proceedings of the 4th International Conference, UC 2005*, Lecture Notes in Computer Science no. 3699, Springer, Heidelberg, 2005.

5. C. S. Calude, M. J. Dinneen, Gh. Păun, G. Rozenberg, and S. Stepney
   (eds.). *Unconventional Computation: Proceedings of the 5th International
   Conference, UC 2006*, Lecture Notes in Computer Science no. 4135, Springer,
   Heidelberg, 2006.

The Steering Committee of the International Conference on Unconventional
Computation series includes T. Bäck (Leiden, The Netherlands), C. S. Calude
(Auckland, New Zealand (Co-chair)), L. K. Grover (Murray Hill, NJ, USA),
J. van Leeuwen (Utrecht, The Netherlands), S. Lloyd (Cambridge, MA, USA),
Gh. Păun (Seville, Spain and Bucharest, Romania), T. Toffoli (Boston, MA,
USA), C. Torras (Barcelona, Spain), G. Rozenberg (Leiden, The Netherlands
and Boulder, Colorado, USA (Co-chair)), and A. Salomaa (Turku, Finland).

The four keynote speakers of the conference for 2007 were:

– Michael A. Arbib (U. Southern California, USA): *A Top-Down Approach to
  Brain-Inspired Computing Architectures*
– Lila Kari (U. Western Ontario, Canada): *Nanocomputing by Self-Assembly*
– Roel Vertegaal (Queen's University, Canada): *Organic User Interfaces (Oui!):
  Designing Computers in Any Way, Shape or Form*
– Tal Mor (Technion–Israel Institute of Technology): *Algorithmic Cooling:
  Putting a New Spin on the Identification of Molecules*

In addition, UC 2007 had two workshops, one on Language Theory in
Biocomputing organized by Michael Domaratzki (University of Manitoba) and
Kai Salomaa (Queen's University), and another on Unconventional Computational Problems, organized by Marius Nagy and Naya Nagy (Queen's University). Moreover, two tutorials were offered on Quantum Information Processing
by Gilles Brassard (Université de Montréal) and Wireless Ad Hoc and Sensor
Networks by Hossam Hassanein (Queen's University).

The Programme Committee is grateful for the much-appreciated work done
by the paper reviewers for the conference. These experts were:

S. G. Akl	G. Franco	N. Nagy	C. Torras
A. G. Barto	M. Hagiya	Gh. Păun	R. Twarock
A. Brabazon	M. Hirvensalo	F. Peper	H. Umeo
C. S. Calude	N. Jonoska	P. H. Potgieter	H. T. Wareham
B. S. Cooper	J. J. Kari	S. Stepney	J. Warren
J. F. Costan	V. Manca	K. Svozil	M. Wilson
M. J. Dinneen	K. Morita	H. Tam	D. Woods
G. Dreyfus	M. Nagy	C. Teuscher	

The Programme Committee consisting of S. G. Akl (Kingston, ON, Canada),
A. G. Barto (Amherst, MA, USA), A. Brabazon (Dublin, Ireland), C. S. Calude
(Auckland, New Zealand), B. S. Cooper (Leeds, UK), J. F. Costa (Lisbon, Por-
tugal), M. J. Dinneen (Auckland, New Zealand (Chair)), G. Dreyfus (Paris,
France), M. Hagiya (Tokyo, Japan), M. Hirvensalo (Turku, Finland), N. Jonoska
(Tampa, FL, USA), J. J. Kari (Turku, Finland), V. Manca (Verona, Italy),

Gh. Păun (Seville, Spain and Bucharest, Romania), F. Peper (Kobe, Japan), P .H. Potgieter (Pretoria, South Africa), S. Stepney (York, UK), K. Svozil (Vienna, Austria), C. Teuscher (Los Alamos, NM, USA), C. Torras (Barcelona, Spain), R. Twarock (York, UK), H. Umeo (Osaka, Japan), H. T. Wareham (St. John's, NL, Canada (Secretary)), and D. Woods (Cork, Ireland) selected 17 papers (out of 27) to be presented as regular contributions.

We extend our thanks to all members of the local Conference Committee, particularly to Selim G. Akl (Chair), Kamrul Islam, Marius Nagy, Yurai Núñez, Kai Salomaa, and Henry Xiao of Queen's University for their invaluable organizational work. We also thank Rhonda Chaytor (St. John's, NL, Canada) for providing additional assistance in preparing the proceedings.

We thank the many local sponsors of the conference.

- Faculty of Arts and Science, Queen's University
- Fields Institute - Research in Mathematical Science
- School of Computing, Queen's University
- Office of Research Services, Queen's University
- Department of Biology, Queen's University
- The Campus Bookstore, Queen's University
- MITACS - Mathematics of Information Technology and Complex Systems
- IEEE, Kingston Section

It is a great pleasure to acknowledge the fine co-operation with the *Lecture Notes in Computer Science* team of Springer for producing this volume in time for the conference.

May 2007

Selim G. Akl
Christian S. Calude
Michael J. Dinneen
Grzegorz Rozenberg
Harold T. Wareham

Table of Contents

Invited Papers

Regular Papers

How Neural Computing Can Still Be Unconventional After All These Years

Michael A. Arbib

USC Brain Project
University of Southern California
Los Angeles, CA USA 90089-2520
arbib@usc.edu

Abstract. Attempts to infer a technology from the computing style of the brain have often focused on general learning styles, such as Hebbian learning, supervised learning, and reinforcement learning. The present talk will place such studies in a broader context based on the diversity of structures in the mammalian brain – not only does the cerebral cortex have many regions with their own distinctive characteristics, but their architecture differs drastically from that of basal ganglia, cerebellum, hippocampus, etc. We will discuss all this within a comparative, evolutionary context. The talk will make the case for a brain-inspired computing architecture which complements the bottom-up design of diverse styles of adaptive subsystem with a top-level design which melds a variety of such subsystems to best match the capability of the integrated system to the demands of a specific range of physical or informational environments.

This talk will be a sequel to *Arbib, M.A., 2003*, Towards a neurally-inspired computer architecture, *Natural computing, 2*:1-46, but the exposition will be self-contained.

S.G. Akl et al.(Eds.): UC 2007, LNCS 4618, p. 1, 2007.
© Springer-Verlag Berlin Heidelberg 2007

Optimal Algorithmic Cooling of Spins

Yuval Elias[1], José M. Fernandez[2], Tal Mor[3], and Yossi Weinstein[4]

[1] Chemistry Department, Technion, Haifa, Israel
[2] Département de génie informatique, École Polytechnique de Montréal, Montréal, Québec, Canada
[3] Computer Science Department, Technion, Haifa, Israel
[4] Physics Department, Technion, Haifa, Israel

Abstract. *Algorithmic Cooling (AC) of Spins* is potentially the first near-future application of quantum computing devices. Straightforward quantum algorithms combined with novel entropy manipulations can result in a method to improve the identification of molecules.

We introduce here several new exhaustive cooling algorithms, such as the Tribonacci and k-bonacci algorithms. In particular, we present the "all-bonacci" algorithm, which appears to reach the maximal degree of cooling obtainable by the optimal AC approach.

1 Introduction

Molecules are built from atoms, and the nucleus inside each atom has a property called "spin". The spin can be understood as the orientation of the nucleus, and when put in a magnetic field, certain spins are binary, either up (ZERO) or down (ONE). Several such bits (inside a single molecule) represent a binary string, or a register. A macroscopic number of such registers/molecules can be manipulated in parallel, as is done, for instance, in Magnetic Resonance Imaging (MRI). The purposes of magnetic resonance methods include the identification of molecules (e.g., proteins), material analysis, and imaging, for chemical or biomedical applications. From the perspective of quantum computation, the spectrometric device that typically monitors and manipulates these bits/spins can be considered a simple "quantum computing" device.

Enhancing the sensitivity of such methods is a Holy Grail in the area of Nuclear Magnetic Resonance (NMR). A common approach to this problem, known as "effective cooling", has been to reduce the entropy of spins. A spin with lower entropy is considered "cooler" and provides a better signal when used for identifying molecules. To date, effective cooling methods have been plagued by various limitations and feasibility problems.

"Algorithmic Cooling" [1,2,3] is a novel and unconventional effective-cooling method that vastly reduces spin entropy. AC makes use of "data compression" algorithms (that are run on the spins themselves) in combination with "thermalization". Due to Shannon's entropy bound (source-coding bound [4]), data compression alone is highly limited in its ability to reduce entropy: the total entropy of the spins in a molecule is preserved, and therefore cooling one spin is done at the expense of heating others. Entropy reduction is boosted dramatically

S.G. Akl et al.(Eds.): UC 2007, LNCS 4618, pp. 2–26, 2007.
© Springer-Verlag Berlin Heidelberg 2007

by taking advantage of the phenomenon of thermalization, the natural return of a spins entropy to its thermal equilibrium value where any information encoded on the spin is erased. Our entropy manipulation steps are designed such that the excess entropy is always placed on pre-selected spins, called "reset bits", which return very quickly to thermal equilibrium. Alternating data compression steps with thermalization of the reset spins thus reduces the total entropy of the spins in the system far beyond Shannon's bound.The AC of short molecules is experimentally feasible in conventional NMR labs; we, for example, recently cooled spins of a three-bit quantum computer beyond Shannon's entropy bound [5].

1.1   Spin-Temperature and NMR Sensitivity

For two-state systems (e.g. binary spins) there is a simple connection between temperature, entropy, and probability. The difference in probability between the two states is called the *polarization bias*. Consider a single spin particle in a constant magnetic field. At equilibrium with a thermal heat-bath the probabilities of this spin to be up or down (i.e., parallel or anti-parallel to the magnetic field) are given by: $p_\uparrow = \frac{1+\varepsilon_0}{2}$, and $p_\downarrow = \frac{1-\varepsilon_0}{2}$. We refer to a spin as a bit, so that $|\uparrow\rangle \equiv |0\rangle$ and $|\downarrow\rangle \equiv |1\rangle$, where $|x\rangle$ represents the spin-state x. The polarization bias is given by $\varepsilon_0 = p_\uparrow - p_\downarrow = \tanh\left(\frac{\hbar\gamma B}{2K_B T}\right)$, where B is the magnetic field, γ is the particle-dependent gyromagnetic constant,[1] K_B is Boltzmann's coefficient, and T is the thermal heat-bath temperature. Let $\varepsilon = \frac{\hbar\gamma B}{2K_B T}$ such that $\varepsilon_0 = \tanh\varepsilon$. For high temperatures or small biases, higher powers of ε can be neglected, so we approximate $\varepsilon_0 \approx \varepsilon$. Typical values of ε_0 for nuclear spins (at room temperature and a magnetic field of ~ 10 Tesla) are $10^{-5} - 10^{-6}$.

A major challenge in the application of NMR techniques is to enhance sensitivity by overcoming difficulties related to the Signal-to-Noise Ratio (SNR). Five fundamental approaches were traditionally suggested for improving the SNR of NMR. Three straightforward approaches - cooling the entire system, increasing the magnetic field, and using a larger sample - are all expensive and limited in applicability, for instance they are incompatible with live samples. Furthermore, such approaches are often impractical due to sample or hardware limitations. A fourth approach - repeated sampling - is very feasible and is often employed in NMR experiments. However, an improvement of the SNR by a factor of M requires M^2 repetitions (each followed by a significant delay to allow relaxation), making this approach time-consuming and overly costly. Furthermore, it is inadequate for samples which evolve over the averaged time-scale, for slow-relaxing spins, or for non-Gaussian noise.

1.2   Effective Cooling of Spins

The fifth fundamental approach to the SNR problem consists of cooling the spins without cooling the environment, an approach known as "effective cooling" of

---

[1] This constant, γ, is thus responsible for the difference in the equilibrium polarization bias of different spins [e.g., a hydrogen nucleus is about 4 times more polarized than a ^{13}C nucleus, but less polarized by three orders of magnitude than an electron spin].

the spins [6,7,8,9]. The effectively-cooled spins can be used for spectroscopy until they relax to thermal equilibrium. The following calculations are done to leading order in ε_0 and are appropriate for $\varepsilon_0 \ll 1$. A spin temperature at equilibrium is $T \propto \varepsilon_0^{-1}$. The single-spin Shannon entropy is $H = 1 - \left(\varepsilon_0^2 / \ln 4 \right)$. A spin temperature out of thermal equilibrium is similarly defined (see for instance [10]). Therefore, increasing the polarization bias of a spin beyond its equilibrium value is equivalent to cooling the spin (without cooling the system) and to decreasing its entropy.

Several more recent approaches are based on the creation of very high polarizations, for example dynamic nuclear polarization [11], para-hydrogen in two-spin systems [12], and hyperpolarized xenon [13]. In addition, there are other spin-cooling methods, based on general unitary transformations [7] and on (closely related) data compression methods in closed systems [8].

One method for effective cooling of spins, *reversible polarization compression (RPC)*, is based on entropy manipulation techniques. RPC can be used to cool some spins while heating others [7,8]. Contrary to conventional data compression,[2] RPC techniques focus on the low-entropy spins, namely those that get colder during the entropy manipulation process. RPC, also termed "molecular-scale heat engine", consists of reversible, in-place, lossless, adiabatic entropy manipulations in a closed system [8]. Therefore, RPC is limited by the second law of thermodynamics, which states that entropy in a closed system cannot decrease, as is also stated by Shannon's source coding theorem [4]. Consider the total entropy of n uncorrelated spins with equal biases, $H(n) \approx n(1 - \varepsilon_0^2 / \ln 4)$. This entropy could be compressed into $m \geq n(1 - \varepsilon^2 / \ln 4)$ high entropy spins, leaving $n - m$ extremely cold spins with entropy near zero. Due to preservation of entropy, the number of extremely cold spins, $n - m$, cannot exceed $n\varepsilon_0^2 / \ln 4$. With a typical $\varepsilon_0 \sim 10^{-5}$, extremely long molecules ($\sim 10^{10}$ atoms) are required in order to cool a single spin to a temperature near zero. If we use smaller molecules, with $n \ll 10^{10}$, and compress the entropy onto $n - 1$ fully-random spins, the entropy of the remaining spin satisfies [2]

$$1 - \varepsilon_{\text{final}}^2 \geq n(1 - \varepsilon_0^2 / \ln 4) - (n - 1) = 1 - n\varepsilon_0^2 / \ln 4. \tag{1}$$

Thus, the polarization bias of the cooled spin is bounded by

$$\varepsilon_{\text{final}} \leq \varepsilon_0 \sqrt{n} . \tag{2}$$

When all operations are unitary, a stricter bound than imposed by entropy conservation was derived by Sørensen [7]. In practice, due to experimental limitations, such as the efficiency of the algorithm, relaxation times, and off-resonance effects, the obtained cooling is significantly below the bound given in eq 2.

Another effective cooling method is known as *polarization transfer (PT)*[6,14]. This technique may be applied if at thermal equilibrium the spins to be used for

---

[2] Compression of data [4] such as bits in a computer file, can be performed by condensation of entropy to a minimal number of high entropy bits, which are then used as a compressed file.

spectroscopy (the observed spins) are less polarized than nearby auxiliary spins. In this case, PT from the auxiliary spins to the observed spins is equivalent to cooling the observed spins (while heating the auxiliary spins). PT from one spin to another is limited as a cooling technique, because the polarization bias increase of the observed spin is bounded by the bias of the highly polarized spin. PT is regularly used in NMR spectroscopy, among nuclear spins on the same molecule [6]. As a simple example, consider the 3-bit molecule trichloroethylene (TCE) shown in Fig. 1. The hydrogen nucleus is about four times more polarized

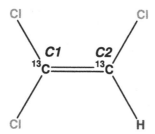

Fig. 1. *A 3-bit computer: a TCE molecule labeled with two* ^{13}C. TCE has three spin nuclei: two carbons and a hydrogen, which are named C1, C2 and H. The chlorines have a very small signal and their coupling with the carbons is averaged out. Therefore, TCE acts as a three-bit computer. The proton can be used as a reset spin because relative to the carbons, its equilibrium bias is four times greater, and its thermalization time is much shorter. Based on the theoretical ideas presented in [2], the hydrogen of TCE was used to cool both carbons, decrease the total entropy of the molecule, and bypass Shannon's bound on cooling via RPC [5].

than each of the carbon nuclei; PT from a hydrogen can be used to cool a single carbon by a factor of four. A different form of PT involves shifting entropy from nuclear spins to electron spins. This technique is still under development [9,13], but has significant potential in the future.

Unfortunately, the manipulation of many spins, say $n > 100$, is a very difficult task, and the gain of $\sqrt{n}$ in polarization is not substantial enough to justify putting this technique into practice.

In its most general form, RPC is applied to spins with different initial polarization biases, thus PT is a special case of RPC. We sometimes refer to both techniques and their combination as *reversible algorithmic cooling*.

2   Algorithmic Cooling

Boykin, Mor, Roychowdhury, Vatan, and Vrijen (hereinafter referred to as BM-RVV), coined the term *Algorithmic Cooling (AC)* for their novel effective-cooling method [1]. AC expands previous effective-cooling techniques by exploiting entropy manipulations in *open systems*. It combines RPC with relaxation (namely, thermalization) of the *hotter spins*, in order to cool far beyond Shannon's entropy bound.

AC employs slow-relaxing spins (which we call *computation spins*) and rapidly relaxing spins (*reset spins*), to cool the system by pumping entropy to the environment. Scheme 1 details the three basic operations of AC. The ratio $R_{relax-times}$, between the spin-lattice relaxation times of the computation spins and the reset spins, must satisfy $R_{relax-times} \gg 1$, to permit the application of many cooling steps to the system.

In all the algorithms presented below, we assume that the relaxation time of each computation spin is sufficiently large, so that the entire algorithm is completed before the computation spins lose their polarization.

The practicable algorithmic cooling (PAC) suggested in [2] indicated a potential for near-future application to NMR spectroscopy [3]. In particular, it presented an algorithm (named PAC2) which uses any odd number of spins such that one of them is a reset spin and the other $2L$ spins are computation spins. PAC2 cools the spins such that the coldest one can (ideally) reach a bias of $(3/2)^L$. This proves an exponential advantage of AC over the best possible reversible algorithmic cooling, as reversible cooling techniques (e.g., of refs [7] and [8]) are limited to a bias improvement factor of $\sqrt{n}$. As PAC2 can be applied to small L (and small n), it is potentially suitable for near future applications.

Scheme 1: *AC is based on the combination of three distinct operations:*

1. *RPC. Reversible Polarization Compression steps redistribute the entropy in the system so that some computation spins are cooled while other computation spins become hotter than the environment.*
2. *SWAP. Controlled interactions allow the hotter computation spins to adiabatically lose their entropy to a set of reset spins, via PT from the reset spins onto these specific computation spins.*
3. *WAIT. The reset spins rapidly thermalize, conveying their entropy to the environment, while the computation spins remain colder, so that the entire system is cooled.*

2.1   Block-Wise Algorithmic Cooling

The original Algorithmic Cooling (BMRVV AC) [1] was designed to address the scaling problem of NMR quantum computing. Thus, a significant number of spins are cooled to a desired level and arranged in a consecutive block, such that the entire block may be viewed as a register of cooled spins. The calculation of the cooling degree attained by the algorithm was based on the law of large numbers, yielding an exponential reduction in spin temperature. Relatively long molecules were required due to this statistical nature, in order to ensure the cooling of 20, 50, or more spins (the algorithm was not analyzed for a small number of spins). All computation spins were assumed to be arranged in a linear chain where each computation spin (e.g., ^{13}C) is attached to a reset spin (e.g., ^{1}H). Reset and computation spins are assumed to have the same bias ε_0. BMRVV AC consists of applying a recursive algorithm repeating (as many times as necessary) the sequence of RPC, SWAP (with reset spins), and WAIT, as detailed in Scheme 1. The relation between gates and NMR pulse sequences is discussed in ref [15].

This cooling algorithm employed a simple form of RPC termed *Basic Compression Subroutine (BCS)* [1]. The computation spins are ordered in pairs, and the following operations are applied to each pair of computation spins X, Y. X_j, Y_j denote the state of the respective spin after stage j; X_0, Y_0 indicate the initial state.

1. Controlled-NOT (CNOT), with spin Y as the control and spin X as target: $Y_1 = Y_0, X_1 = X_0 \oplus Y_0$, where $\oplus$ denotes exclusive OR (namely, addition modulo 2). This means that the target spin is flipped (NOT gate $|\uparrow\rangle \leftrightarrow |\downarrow\rangle$) if the control spin is $|1\rangle$ (i.e., $|\downarrow\rangle$). Note that $X_1 = |0\rangle \Leftrightarrow X_0 = Y_0$, in which case the bias of spin Y is doubled, namely the probability that $Y_1 = |0\rangle$ is very close to $(1 + 2\varepsilon_0)/2$.
2. NOT gate on spin X_1. So $X_2 = NOT(X_1)$.
3. Controlled-SWAP (CSWAP), with spin X as a control: if $X_2 = |1\rangle$ (Y was cooled), transfer the improved state of Y (doubled bias) to a chosen location by alternate SWAP and CSWAP gates.

Let us show that indeed in step 1 the bias of spin Y is doubled whenever $X_1 = |0\rangle$. Consider the truth table for the CNOT operation below:

$$
\begin{array}{ccc}
input : X_0 Y_0 & output : X_1 Y_1 \\
0\ 0 & \rightarrow & 0\ 0 \\
0\ 1 & \rightarrow & 1\ 1 \\
1\ 0 & \rightarrow & 1\ 0 \\
1\ 1 & \rightarrow & 0\ 1
\end{array}
$$

The probability that both spins are initially $|0\rangle$ is $p_{00} \equiv P(X_0 = |0\rangle, Y_0 = |0\rangle) = (1 + \varepsilon_0)^2/4$. In general, $p_{kl} \equiv P(X_0 = |k\rangle, Y_0 = |l\rangle)$ such that $p_{01} = p_{10} = (1+\varepsilon_0)(1-\varepsilon_0)/4$, and $p_{11} = (1-\varepsilon_0)^2/4$. After CNOT, the conditional probability $P(Y_1 = |0\rangle | X_1 = |0\rangle)$ is $\frac{q_{00}}{q_{00}+q_{01}} = \frac{p_{00}}{p_{00}+p_{11}}$, where $q_{kl} \equiv P(X_1 = |k\rangle, Y_1 = |l\rangle)$ and q_{kl} is derived from p_{kl} according to the truth table, so that $q_{00} = p_{00}$, and $q_{01} = p_{11}$. In terms of ε_0 this probability is

$$
\frac{(1 + \varepsilon_0)^2}{(1 + \varepsilon_0)^2 + (1 - \varepsilon_0)^2} \approx \frac{1 + 2\varepsilon_0}{2},
$$

indicating that the bias of Y was indeed doubled in this case.

For further details regarding the basic compression subroutine and its application to many pairs of spins (in a recursive manner) to reach a bias of $2^j \varepsilon_0$, we refer the reader to ref [1].

2.2 Practicable Algorithmic Cooling (PAC)

An efficient and experimentally feasible AC technique was later presented, termed *"practicable algorithmic cooling (PAC)"* [2]. Unlike the first algorithm, the analysis of PAC does not rely on the law of large numbers, and no statistical requirement is invoked. PAC is thus a simple algorithm that may be conveniently analyzed, and which is already applicable for molecules containing very few spins. Therefore, PAC has already led to experimental implementations. PAC algorithms use PT steps, reset steps, and *3-bit-compression (3B-Comp)*. As already mentioned, one of the algorithms presented in [2], PAC2, cools the spins such that the coldest one can (ideally) reach a bias of $(3/2)^L$, while the number of spins is only $2L + 1$. PAC is simple, as all compressions are applied to three spins, often with identical biases. The algorithms we present in the next section are more efficient and lead to a better degree of cooling, but are also more complicated. We believe that PAC is the best candidate for near future applications of AC, such as improving the SNR of biomedical applications.

PAC algorithms [2] use a basic 3-spin RPC step termed 3-bit-compression (3B-Comp):

Scheme 2: *3-BitCompression (3BComp)*

1. *CNOT, with spin B as a control and spin A as a target. Spin A is flipped if $B = |1\rangle$.*
2. *NOT on spin A.*
3. *CSWAP with spin A as a control and spins B and C as targets. B and C are swapped if $A = |1\rangle$.*

Assume that the initial bias of the spins is ε_0. The result of scheme 2 is that spin C is cooled: if $A = |1\rangle$ after the first step (and $A = |0\rangle$ after the second step), C is left unchanged (with its original bias ε_0); if however, $A = |0\rangle$ after the first step (hence $A = |1\rangle$ after the second step), spin B is cooled by a factor of about 2 (see previous subsection), and following the CSWAP the new bias is placed on C. Therefore, on average, C is cooled by a factor of $3/2$. We do not care about the biases of the other two spins, as they subsequently undergo a reset operation.

In many realistic cases the polarization bias of the reset spins at thermal equilibrium, ε_0, is higher than the biases of the computation spins. Thus, an initial PT from reset spins to computation spins (e.g., from hydrogen to carbon or nitrogen), cools the computation spins to the 0th purification level, ε_0.

As an alternative to Scheme 2, the 3B-Comp operation depicted in Scheme 3 is similar to the CNOT-CSWAP combination (Scheme 2) and cools spin C to the same degree. This gate is known as the MAJORITY gate since the resulting value of bit C indicates whether the majority of the bits had values of $|0\rangle$ or $|1\rangle$ prior to the operation of the gate.

Scheme 3: *Single operation implementing 3B-Comp*
Exchange the states $|100\rangle \leftrightarrow |011\rangle$.
Leave the rest of the states unchanged.

If 3B-Comp is applied to three spins {C,B,A} which have identical biases, $\varepsilon_C = \varepsilon_B = \varepsilon_A = \varepsilon_0$, then spin C will acquire a new bias ε'_C. This new bias is obtained from the initial probability that spin C is $|0\rangle$ by adding the effect of exchanging $|100\rangle \leftrightarrow |011\rangle$:

$$\frac{1+\varepsilon'_C}{2} = \frac{1+\varepsilon_C}{2} + p_{|100\rangle} - p_{|011\rangle} \tag{3}$$

$$= \frac{1+\varepsilon_0}{2} + \frac{1-\varepsilon_0}{2}\frac{1+\varepsilon_0}{2}\frac{1+\varepsilon_0}{2} - \frac{1+\varepsilon_0}{2}\frac{1-\varepsilon_0}{2}\frac{1-\varepsilon_0}{2} = \frac{1 + \frac{3\varepsilon_0 - \varepsilon_0^3}{2}}{2}.$$

The resulting bias is

$$\varepsilon'_C = \frac{3\varepsilon_0 - \varepsilon_0^3}{2}, \tag{4}$$

and in the case where $\varepsilon_0 \ll 1$

$$\varepsilon'_C \approx \frac{3\varepsilon_0}{2}. \tag{5}$$

We have reviewed two schemes for cooling spin C: one using CNOT and CSWAP gates (see Scheme 2), and the other using the MAJORITY gate (see Scheme 3). Spin C reaches the same final bias following both schemes. The other spins may obtain different biases, but this is irrelevant for our purpose, as they undergo a reset operation in the next stages of any cooling algorithm.

The simplest practicable cooling algorithm, termed Practicable Algorithmic Cooling 1 (PAC1) [2], employs dedicated reset spins, which are not involved in compression steps. PAC1 on three computation spins is shown in Fig. 2, where each computation spin X has a neighboring reset spin, r_X, with which it can be swapped. The following examples illustrate cooling by PAC1. In order to cool a single spin (say, spin C) to the first purification level, start with three computation spins, CBA, and perform the sequence presented in Example 1. [3]

Example 1: *Cooling spin C to the 1st purification level by PAC1*

1. *$PT((r_C \to C); (r_B \to B); (r_A \to A))$, to initiate all spins.*
2. *3B-Comp$(C; B; A)$, increase the polarization of C.*

A similar algorithm can also cool the entire molecule beyond Shannon's entropy bound. See the sequence presented in Example 2.

Example 2: *Cooling spin C and bypassing Shannon's entropy bound by PAC1*

1. *$PT((r_C \to C); (r_B \to B); (r_A \to A))$, to initiate all spins.*
2. *3B-Comp$(C; B; A)$, increase the polarization of C.*
3. *WAIT*
4. *$PT((r_B \to B); (r_A \to A))$, to reset spins A, B.*

In order to cool one spin (say, spin E) to the second purification level (polarization bias ε_2), start with five computation spins ($EDCBA$) and perform the sequence presented in Example 3.

For small biases, the polarization of spin E following the sequence in Example 3 is $\varepsilon_2 \approx (3/2)\varepsilon_1 \approx (3/2)^2\varepsilon_0$.

---

[3] This sequence may be implemented by the application of an appropriate sequence of radiofrequency pulses.

Example 3: *Cooling spin E in $EDCBA$ to the* 2nd *purification level by PAC1*

1. *$PT(E;D;C)$, to initiate spins EDC.*
2. *$3B\text{-}Comp(E;D;C)$, increase the polarization of E to ε_1.*
3. *WAIT*
4. *$PT(D;C;B)$, to initiate spins DCB.*
5. *$3B\text{-}Comp(D;C;B)$, increase the polarization of D to ε_1.*
6. *WAIT*
7. *$PT(C;B;A)$, to initiate spins CBA.*
8. *$3B\text{-}Comp(C;B;A)$, increase the polarization of C to ε_1.*
9. *$3B\text{-}Comp(E;D;C)$, increase the polarization of E to ε_2.*

For molecules with more spins, a higher cooling level can be obtained.

This simple practicable cooling algorithm (PAC1) is easily generalized to cool one spin to any purification level L. [2] The resultant bias will be very close to $(3/2)^L$, as long as this bias is much smaller than 1. For a final bias that approaches 1 (close to a pure state), as required for conventional (non-ensemble) quantum computing, a more precise calculation is required.

Consider an array of n computation spins, $c_n c_{n-1} \ldots c_2 c_1$, where each computation spin, c_i, is attached to a reset spin, r_i (see Fig. 2 for the case of $n = 3$). To cool c_k, the spin at index k, to a purification level $j \in 1 \ldots L$ the procedure $M_j(k)$ was recursively defined as follows [2]: $M_0(k)$ is defined as a single PT step from reset spin r_k to computation spin c_k to yield a polarization bias of ε_0 (the 0th purification level). The procedure $M_1(k)$ applies M_0 to the three

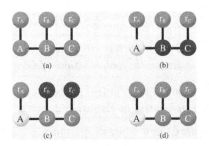

Fig. 2. An abstract example of a molecule with three computation spins A, B and C, attached to reset spins r_A, r_B, and r_C, respectively. All spins have the same equilibrium polarization bias, ε_0 (a). The temperature after each step is illustrated by color: gray - thermal equilibrium; white - colder than initial temperature; and black - hotter than initial temperature. PAC uses 3-bit compression (3B-Comp), Polarization Transfer (PT) steps, and RESET steps: 1. $3B\text{-}Comp(C;B;A)$; the outcome of this step is shown in (b). 2. $PT(r_B \rightarrow B)$, $PT(r_C \rightarrow C)$; the outcome is shown in (c). 3. $RESET(r_B, r_C)$; the outcome is shown in (d). The 3-bit-compression applied in the first step operates on the three computation spins, increasing the bias of spin A by a factor of $3/2$, while heating the other two spins. The 3B-Comp step cools spin A, and the following PT and RESET steps restore the initial biases of the other spins, thus the entire system is cooled.

spins followed by 3B-Comp on these spins, so that spin c_k is cooled to the first purification level. Similarly, $M_2(k)$ applies M_1 three times to cool $c_k; c_{k-1}; c_{k-2}$ to the first purification level, followed by 3B-Comp on these spins, so that spin c_k is cooled to the second purification level. We use the notation $\mathcal{B}_{(j-1)\to j}(k)$ to represent the application of 3B-Comp to spins to purify spin c_k from ε_{j-1} to ε_j. Then, the full algorithm has a simple recursive form, described in Algorithm 1.

Algorithm 1: *Practicable algorithmic cooling 1 (PAC1):*
For $j \in \{1, \ldots, L\}$

$$M_j(k) = \mathcal{B}_{\{(j-1)\longrightarrow j\}}(k)M_{j-1}(k-2)\, M_{j-1}(k-1)\, M_{j-1}(k)\,, \qquad (6)$$

applied from right to left ($M_{j-1}(k)$ is applied first).

For instance, $M_1(3) = \mathcal{B}_{\{0\to 1\}}(3)M_0(1)\, M_0(2)\, M_0(3)$, is 3B-Comp applied after reset as described in Example 1.[4] Clearly, $M_1(k)$ can be applied to any $k \geq 3$, $M_2(k)$ to $k \geq 5$, and $M_j(k)$ to $k \geq 2j + 1$. Thus, to cool a single spin to a purification level of L, $2L + 1$ computation spins and an equal number of reset spins are required. A single reset spin could be used for initializing all relevant computation spins, at the expense of additional time steps.

Reset spins may also be used for compression, thus replacing the 3B-Comp and PT steps above by a generalized RPC. The corresponding algorithm, termed PAC2, has an improved space complexity relative to PAC1. We explicitly show how this is achieved. Let ε_0 be the polarization bias of the reset spin. In order to cool a single spin to ε_1, start with two computation spins, CB, and one reset spin, A, and perform the sequence shown in Example 4 to cool spin C.

Example 4: *Cooling spin C to the 1ˢᵗ purification level by PAC2*

1. *PT(A → B).*
2. *PT(B → C) to initiate spin C.*
3. *RESET(A) (by waiting).*
4. *PT(A → B) to initiate spin B.*
5. *RESET(A). If the thermalization time of the computation spins is sufficiently large, there are now three spins with polarization bias ε_0.*
6. *3B-Comp to increase the polarization of spin C to ε_1.*

In order to cool one spin (say, spin E) to the second purification level (polarization bias ε_2), start with 5 computation spins ($EDCBA$) and follow Example 5.

Example 5: *Cooling spin E in EDCBA to the 2ⁿᵈ purification level by PAC2*

1. *PT sequentially to initiate spins EDC (RESET(A) after each PT).*
2. *3B-Comp on spins EDC to increase the polarization of spin E to ε_1.*
3. *PT sequentially to initiate spins DCB (RESET(A) after each PT).*
4. *3B-Comp on spins DCB to increase the polarization of spin D to ε_1.*
5. *PT sequentially to initiate spins CB (RESET(A) after each PT).*

---

[4] The procedure of cooling one spin to the second level (starting with five spins) is written as $M_2(5) = \mathcal{B}_{\{1\to 2\}}(5)M_1(3)\, M_1(4)\, M_1(5)$.

6. *3B-Comp on spins CBA to increase the polarization of spin C to ε_1.*
7. *3B-Comp on spins EDC to increase the polarization of spin E to ε_2.*

By repeated application of PAC2 in a recursive manner (as for PAC1), spin systems can be cooled to very low temperatures. PAC1 uses dedicated reset spins, while PAC2 also employs reset spins for compression. The simplest algorithmic cooling can thus be obtained with as few as 3 spins, comprising 2 computation spins and one reset spin.

The algorithms presented so far applied compression steps (3B-Comp) to three identical biases (ε_0); recall that this cools one spin to a new bias, $\varepsilon'_C \approx (3/2)\varepsilon_0$. Now consider applying compression to three spins with different biases ($\varepsilon_C, \varepsilon_B, \varepsilon_A$); spin C will acquire a new bias ε'_C, which is a function of the three initial biases [16,17]. This new bias is obtained from the initial probability that spin C is $|0\rangle$ by adding the effect of the exchange $|100\rangle \leftrightarrow |011\rangle$:

$$
\begin{aligned}
\frac{1+\varepsilon'_C}{2} &= \frac{1+\varepsilon_C}{2} + p_{100} - p_{011} \\
&= \frac{1+\varepsilon_C}{2} + \frac{1-\varepsilon_C}{2}\frac{1+\varepsilon_B}{2}\frac{1+\varepsilon_A}{2} - \frac{1+\varepsilon_C}{2}\frac{1-\varepsilon_B}{2}\frac{1-\varepsilon_A}{2} \\
&= \frac{1 + \frac{\varepsilon_C+\varepsilon_B+\varepsilon_A-\varepsilon_C\varepsilon_B\varepsilon_A}{2}}{2}.
\end{aligned}
\tag{7}
$$

The resulting bias is

$$
\varepsilon'_C = \frac{\varepsilon_C + \varepsilon_B + \varepsilon_A - \varepsilon_C\varepsilon_B\varepsilon_A}{2},
\tag{8}
$$

and in the case where $\varepsilon_C, \varepsilon_B, \varepsilon_A \ll 1$,

$$
\varepsilon'_C \approx \frac{\varepsilon_C + \varepsilon_B + \varepsilon_A}{2}.
\tag{9}
$$

3   Exhaustive Cooling Algorithms

The following examples and derivations are to leading order in the biases. This is justified as long as all biases are much smaller than 1, including the final biases. For example, with $\varepsilon_0 \sim 10^{-5}$ (ε_0 is in the order of magnitude of 10^{-5}) and $n \leq 13$, the calculations are fine (see section 4 for details).

3.1   Exhaustive Cooling on Three Spins

Example 6: *Fernandez [16]:* $\mathcal{F}(C, B, A, m)$
Repeat the following m times

1. *3B-Comp$(C; B; A)$, to cool C.*
2. *RESET$(B; A)$*

Consider an application of the primitive algorithm in Example 6 to three spins, C, B, A, where A and B are reset spins with initial biases of $\varepsilon_A = \varepsilon_B = \varepsilon_0$ and

$\varepsilon_C^{(0)} = 0$ (the index over the bias of C denotes the iteration). After each WAIT step, spins A and B are reset back to their equilibrium bias, ε_0. As B and A play roles of both reset and computation spins, the operation RESET in Example 6 simply means WAIT. From eq 9, after the first iteration

$$\varepsilon_C^{(1)} = \frac{\varepsilon_C^{(0)} + \varepsilon_B + \varepsilon_A}{2} = \frac{0 + 2\varepsilon_0}{2} = \varepsilon_0.$$

After the second iteration

$$\varepsilon_C^{(2)} = \frac{\varepsilon_C^{(1)} + \varepsilon_B + \varepsilon_A}{2} = \frac{\varepsilon_0 + 2\varepsilon_0}{2} = \frac{3\varepsilon_0}{2}.$$

After the mth iteration

$$\varepsilon_C^{(m)} = \frac{\varepsilon_C^{(m-1)} + 2\varepsilon_0}{2} = 2^{-m}\varepsilon_C^{(0)} + 2\varepsilon_0 \sum_{j=1}^{m} 2^{-j} = 0 + \left(1 - 2^{-m}\right) 2\varepsilon_0. \tag{10}$$

The asymptotic bias ($m \to \infty$) may be extracted from

$$\varepsilon_C^{(m)} \approx \frac{\varepsilon_C^{(m)} + 2\varepsilon_0}{2}, \tag{11}$$

with the unique solution and bias configuration

$$\varepsilon_C^{(m)} = 2\varepsilon_0 \Rightarrow \{2\varepsilon_0, \varepsilon_0, \varepsilon_0\}. \tag{12}$$

In order to achieve good asymptotics, one would like to reach $\varepsilon_C = (2 - \delta)\varepsilon_0$, where δ is arbitrarily small. In this case the number of iterations required is given by $2^{1-m} = \delta \implies m = 1 + \lceil \log_2(1/\delta) \rceil$. For example, if $\delta = 10^{-5}$, 18 repetitions are sufficient. Up to an accuracy of δ, the biases after the last reset are as in eq 12.

3.2   The Fibonacci Algorithm

An algorithm based on 3B-Comp was recently devised [18], which produces a bias configuration that asymptotically approaches the Fibonacci series. In particular, when applied to n spins, the coldest spin attains the bias $\varepsilon_{\text{final}} \approx \varepsilon_0 F_n$, where F_n is the nth element of the series and $\varepsilon_0 F_n \ll 1$. Note that $\varepsilon_C^{(m)}$ of eq 12 is $\varepsilon_0 F_3$. Also note that for 12 spins and $\varepsilon_0 \sim 10^{-5}$, $\varepsilon_0 F_{12} \ll 1$, so the approximation is useful for non-trivial spin systems. Compare the bias enhancement factor in this case, $F_{12} = 144$, to PAC2 with 13 spins - $(3/2)^6 \approx 11$.

Example 7 expands Example 6 to four spins with initial biases $\varepsilon_D = \varepsilon_C = 0$ and $\varepsilon_B = \varepsilon_A = \varepsilon_0$ (A, B are reset spins). We refer to the parameter m from Example 6 as m_3, the number of repetitions applied to three spins. We refer to $\mathcal{F}$ from Example 6 as $\mathcal{F}_2$ for consistency with the rest of the section. Consider the following example:

Example 7: $\mathcal{F}_2(D, C, B, A, m_4, m_3)$
Repeat the following m_4 times:

1. *3B-Comp(D; C; B), places the new bias on spin D.*
2. *$\mathcal{F}_2(C, B, A, m_3)$.*

After each iteration, i, the bias configuration is $\{\varepsilon_D^{(i)}, \varepsilon_C^{m_3}, \varepsilon_0, \varepsilon_0\}$. Running the two steps in Example 7 exhaustively ($m_4, m_3 \gg 1$) yields:

$$\varepsilon_C^{(m_3)} \approx 2\varepsilon_0,$$

$$\varepsilon_D^{(m_4)} \approx \frac{\varepsilon_B + \varepsilon_C^{(m_3)} + \varepsilon_D^{(m_4)}}{2} \tag{13}$$

$$\Rightarrow \varepsilon_D^{(m_4)} = \varepsilon_B + \varepsilon_C^{(m_3)} = 3\varepsilon_0 = \varepsilon_0 F_4.$$

This unique solution was obtained by following the logic of eqs 11 and 12.

We generalize this algorithm to n spins $A_n, \ldots, A_1$ in Algorithm 2.

Algorithm 2: *Fibonacci $\mathcal{F}_2(A_n, \ldots, A_1, m_n, , m_3)$*
Repeat the following m_n times:

1. *3B-Comp(A_n; A_{n-1}; A_{n-2}).*
2. *$\mathcal{F}_2(A_{n-1}, \ldots, A_1, m_{n-1}, \ldots, m_3)$.*

[with $\mathcal{F}_2(A_3, A_2, A_1, m_3)$ defined by Example 6.]

Note that different values of $m_{n-1}, \ldots, m_3$ may be applied at each repetition of step 2 in Algorithm 2. This is a recursive algorithm; it calls itself with one less spin. Running Algorithm 2 exhaustively ($m_n, m_{n-1}, \ldots, m_3 \gg 1$) results, similarly to eq 13, in

$$\varepsilon_{A_n}^{(m_n)} \approx \frac{\varepsilon_{A_n}^{(m_n)} + \varepsilon_{A_{n-1}} + \varepsilon_{A_{n-2}}}{2}$$

$$\Rightarrow \varepsilon_{A_n}^{(m_n)} \approx \varepsilon_{A_{n-1}} + \varepsilon_{A_{n-2}}. \tag{14}$$

This formula yields the Fibonacci series $\{\ldots, 8, 5, 3, 2, 1, 1\}$, therefore $\varepsilon_{A_i} \rightarrow \varepsilon_0 F_i$. We next devise generalized algorithms which achieve better cooling. An analysis of the time requirements of the Fibonacci cooling algorithm is provided in [18].

3.3   The Tribonacci Algorithm

Consider 4-bit-compression (4B-Comp) which consists of an exchange between the states $|1000\rangle$ and $|0111\rangle$ (the other states remain invariant similar to 3B-Comp). Application of 4B-Comp to four spins D, C, B, A with corresponding biases $\varepsilon_D, \varepsilon_C, \varepsilon_B, \varepsilon_A \ll 1$ results in a spin with the probability of the state $|0\rangle$ given by

$$\frac{1 + \varepsilon_D'}{2} = \frac{1 + \varepsilon_D}{2} + p_{|1000\rangle} - p_{|0111\rangle} \approx \frac{1 + \frac{\varepsilon_A + \varepsilon_B + \varepsilon_C + 3\varepsilon_D}{4}}{2}, \tag{15}$$

following the logic of eqs 7 and 8, and finally

$$\varepsilon_D' \approx (\varepsilon_A + \varepsilon_B + \varepsilon_C + 3\varepsilon_D)/4. \tag{16}$$

Example 8 applies an algorithm based on 4B-Comp to 4 spins, with initial biases $\varepsilon_D = \varepsilon_C = 0$ (A, B are reset spins). In every iteration of Example 8, running step 2 exhaustively yields the following biases: $\varepsilon_C = 2\varepsilon_0, \varepsilon_B = \varepsilon_A = \varepsilon_0$. The compression step (step 1) is then applied onto the configuration

Example 8: $\mathcal{F}_3(D, C, B, A, m_4, m_3)$
Repeat the following m_4 times:

1. *4B-Comp(D;C;B;A).*
2. $\mathcal{F}_2(C, B, A, m_3)$.

From eq 16, $\varepsilon_D^{(i+1)} = (4\varepsilon_0 + 3\varepsilon_D^{(i)})/4$. For sufficiently large m_4 and m_3 the algorithm produces final polarizations of

$$\varepsilon_D^{(m_4)} \approx \varepsilon_0 + \frac{3}{4}\varepsilon_D^{(m_4)} \implies \varepsilon_D^{(m_4)} \approx 4\varepsilon_0. \tag{17}$$

For more than 4 spins, a similar 4B-Comp based algorithm may be defined. Example 9 applies an algorithm based on 4B-Comp to 5 spins.

Example 9: $\mathcal{F}_3(E, D, C, B, A, m_5, m_4, m_3)$
Repeat the following m_5 times:

1. *4B-Comp(E; D; C; B).*
2. $\mathcal{F}_3(D, C, B, A, m_4, m_3)$.

Step 2 is run exhaustively in each iteration; the biases of $DCBA$ after this step are $\varepsilon_D = 4\varepsilon_0, \varepsilon_C = 2\varepsilon_0, B = A = \varepsilon_0$. The 4B-Comp step is then applied to the biases of spins (E, D, C, B). Similarly to eq 16, $\varepsilon_E^{(i+1)} = (7\varepsilon_0 + 3\varepsilon_E^{(i)})/4$. Hence, for sufficiently large m_5 and m_4 the final bias of E is

$$\varepsilon_E^{(m_5)} \approx (7\varepsilon_0 + 3\varepsilon_E^{(m_5)})/4 \Rightarrow \varepsilon_E^{(m_5)} \approx 7\varepsilon_0. \tag{18}$$

Algorithm 3 applies to an arbitrary number of spins $n > 4$. This is a recursive algorithm, which calls itself with one less spin.

Algorithm 3: *Tribonacci:* $\mathcal{F}_3(A_n, \ldots, A_1, m_n, \ldots, m_3)$
Repeat the following m_n times:

1. *4B-Comp(A_n; A_{n-1}; A_{n-2}; A_{n-3}).*
2. $\mathcal{F}_3(A_{n-1}, \ldots, A_1, m_{n-1}, \ldots, m_3)$.

[With $\mathcal{F}_3(A_4, A_3, A_2, A_1, m_4, m_3)$ given by Example 8.]
The compression step, 4B-Comp, is applied to

$$\varepsilon_{A_n}^{(i)}, \varepsilon_{A_{n-1}}, \varepsilon_{A_{n-2}}, \varepsilon_{A_{n-3}},$$

and results in

$$\varepsilon_{A_n}^{(i+1)} = (\varepsilon_{A_{n-1}} + \varepsilon_{A_{n-2}} + \varepsilon_{A_{n-3}} + 3\varepsilon_{A_n}^{(i)})/4. \tag{19}$$

For sufficiently large $m_j, j = 3, \ldots, n$

$$\varepsilon_{A_n}^{(m_n)} \approx (\varepsilon_{A_{n-1}} + \varepsilon_{A_{n-2}} + \varepsilon_{A_{n-3}} + 3\varepsilon_{A_n}^{(m_n)})/4$$

$$\Rightarrow \varepsilon_{A_n}^{(m_n)} \approx \varepsilon_{A_{n-1}} + \varepsilon_{A_{n-2}} + \varepsilon_{A_{n-3}}. \tag{20}$$

For $\varepsilon_{A_3} = 2\varepsilon_0, \varepsilon_{A_2} = \varepsilon_{A_1} = \varepsilon_0$, the resulting bias will be $\varepsilon_{A_n}^{(m_n)} \approx \varepsilon_0 T_n$, where T_n is the nth Tribonacci number.[5] As for the Fibonacci algorithm, we assume $\varepsilon_0 T_n \ll 1$. The resulting series is $\{\ldots, 24, 13, 7, 4, 2, 1, 1\}$.

3.4   The k-Bonacci Algorithm

A direct generalization of the Fibonacci and Tribonacci algorithms above is achieved by the application of $(k + 1)$-bit-compression, or $(k+1)$B-Comp. This compression on $k+1$ spins involves the exchange of $|100\cdots000\rangle$ and $|011\cdots111\rangle$, leaving the other states unchanged. When $(k + 1)$B-Comp is applied to $k + 1$ spins with biases $\{\varepsilon_{A_{k+1}}, \varepsilon_{A_k}, \ldots, \varepsilon_{A_2}, \varepsilon_{A_1}\}$, where A_1 and A_2 are reset spins, the probability that the leftmost spin is $|0\rangle$ becomes (similarly to eq 15)

$$\frac{1 + \varepsilon'_{k+1}}{2} = \frac{1 + \varepsilon_{k+1}}{2} + p_{100\cdots000} - p_{011\cdots111} \approx \frac{1 + \frac{(2^{k-1}-1)\varepsilon_{A_{k+1}} + \sum_{j=1}^{k} \varepsilon_{A_j}}{2^{k-1}}}{2}. \tag{21}$$

Therefore, the bias of the leftmost spin becomes

$$\varepsilon'_{k+1} \approx \frac{(2^{k-1} - 1)\varepsilon_{A_{k+1}} + \sum_{j=1}^{k} \varepsilon_{A_j}}{2^{k-1}}. \tag{22}$$

Example 10 applies an algorithm based on $(k + 1)$B-Comp to $k + 1$ spins. $\mathcal{F}_k$ on $k + 1$ spins calls (recursively) $\mathcal{F}_k - 1$ on k spins (recall that F_3 on four spins called F_2 on three spins).

Example 10: $\mathcal{F}_k(A_{k+1}, \ldots, A_1, m_{k+1}, \ldots, m_3)$
repeat the following m_{k+1} times:

1. $(k + 1)$B-Comp$(A_{k+1}, A_k, \ldots, A_2, A_1)$
2. $\mathcal{F}_{k-1}(A_k, \ldots, A_1, m_k, \ldots, m_3)$

If step 2 is run exhaustively at each iteration, the resulting biases are

$$\varepsilon_{A_k} = 2^{k-2}\varepsilon_0, \varepsilon_{A_{k-1}} = 2^{k-3}\varepsilon_0, \ldots, \varepsilon_{A_3} = 2\varepsilon_0, \varepsilon_{A_2} = \varepsilon_{A_1} = \varepsilon_0.$$

The $(k + 1)$B-Comp step is then applied to the biases $\varepsilon_{A_{k+1}}^{(i)}, 2^{k-2}\varepsilon_0, 2^{k-3}\varepsilon_0, \ldots,$ $2\varepsilon_0, \varepsilon_0, \varepsilon_0$. From eq 22,

$$\varepsilon_{A_{k+1}}^{(i+1)} \approx \frac{(2^{k-1} - 1)\varepsilon_{A_{k+1}}^{(i)} + 2^{k-1}\varepsilon_0}{2^{k-1}}.$$

---

[5] The Tribonacci series (also known as the Fibonacci 3-step series) is generated by the recursive formula $a_i = a_{i-1} + a_{i-2} + a_{i-3}$, where $a_3 = 2$ and $a_2 = a_1 = 1$.

Hence, for sufficiently large $m_j, j = 3, \ldots, k+1$, the final bias of A_{k+1} is

$$\varepsilon_{A_{k+1}}^{(m_{k+1})} \approx \frac{(2^{k-1}-1)\varepsilon_{A_{k+1}}^{(m_{k+1})} + 2^{k-1}\varepsilon_0}{2^{k-1}} \Rightarrow \varepsilon_{A_{k+1}}^{(m_{k+1})} \approx 2^{k-1}\varepsilon_0. \qquad (23)$$

For more than $k+1$ spins a similar $(k+1)$B-Comp based algorithm may be defined. Example 11 applies such an algorithm to $k+2$ spins.

Example 11: $\mathcal{F}_k(A_{k+2}, \ldots, A_1, m_{k+2}, \ldots, m_3)$
Repeat the following steps m_{k+2} times:

1. $(k+1)$B-Comp$(A_{k+2}, A_{k+1}, \ldots, A_3, A_2)$
2. $\mathcal{F}_k(A_{k+1}, \ldots, A_1, m_{k+1}, \ldots, m_3)$ *[defined in Example 11.]*

When step 2 is run exhaustively at each iteration, the resulting biases are

$$\varepsilon_{A_{k+1}} = 2^{k-1}\varepsilon_0, \varepsilon_{A_k} = 2^{k-2}\varepsilon_0, \ldots, \varepsilon_{A_3} = 2\varepsilon_0, \varepsilon_{A_2} = \varepsilon_{A_1} = \varepsilon_0.$$

The $(k+1)$B-Comp is applied to the biases. From Eq. 22,

$$\varepsilon_{A_{k+2}}^{(i+1)} = \frac{(2^{k-1}-1)\varepsilon_{A_{k+2}}^{(i)} + (2^k-1)\varepsilon_0}{2^{k-1}}.$$

Hence, for sufficiently large $m_j, j = 3, \ldots, k+2$, the final bias of A_{k+2} is

$$\varepsilon_{A_{k+2}}^{(m_{k+2})} \approx \frac{(2^{k-1}-1)\varepsilon_{A_{k+2}}^{(m_{k+2})} + (2^k-1)\varepsilon_0}{2^{k-1}} \Rightarrow \varepsilon_{A_{k+2}}^{(m_{k+2})} \approx (2^k-1)\varepsilon_0. \qquad (24)$$

Algorithm 4 generalizes Examples 10 and 11.

Algorithm 4: k-bonacci: $\mathcal{F}_k(A_n, \ldots, A_1, m_n, \ldots, m_3)$
Repeat the following m_n times:

1. $(k+1)$B-Comp$(A_n, A_{n-1}, \ldots, A_{n-k})$.
2. $\mathcal{F}_k(A_{n-1}, \ldots, A_1, m_{n-1}, \ldots, m_3)$.

[with $\mathcal{F}_k(A_{k+1}, \ldots, A_1, m_{k+1}, \ldots, m_3)$ defined in Example 10.]

The algorithm is recursive; it calls itself with one less spin. The compression step, $(k+1)$B-Comp, is applied to

$$\varepsilon_{A_n}^{(i)}, \varepsilon_{A_{n-1}}, \ldots \varepsilon_{A_{n-k}}.$$

From Eq. 22, the compression results in

$$\varepsilon_{A_n}^{(i+1)} \approx \frac{(2^{k-1}-1)\varepsilon_{A_n}^{(i)} + \sum_{j=1}^{k}\varepsilon_{A_{n-j}}}{2^{k-1}}. \qquad (25)$$

For sufficiently large m_n, m_{n-1}, etc.

$$\varepsilon_{A_n}^{(m_n)} \approx \frac{(2^{k-1}-1)\varepsilon_{A_n}^{(m_n)} + \sum_{j=1}^{k}\varepsilon_{A_{n-j}}}{2^{k-1}}$$

$$\Rightarrow \varepsilon_{A_n}^{(m_n)} \approx \sum_{j=1}^{k}\varepsilon_{A_{n-j}} \qquad (26)$$

This set of biases corresponds to the k-step Fibonacci sequence which is generated by a recursive formula.

$$a_1, a_2 = 1, a_\ell = \left\{ \begin{array}{c} \sum_{i=1}^{\ell-1} a_{\ell-i}, \; 3 \le \ell \le k+1 \\ \\ \sum_{i=1}^{k} a_{\ell-i} \; \ell > k+1 \end{array} \right\}. \tag{27}$$

Notice that for $3 \le \ell \le k+1$,

$$a_\ell = \sum_{i=1}^{\ell-1} a_i = 1 + 1 + 2 + 4 + \cdots + 2^{\ell-4} + 2^{\ell-3} = 2^{\ell-2}. \tag{28}$$

The algorithm uses ℓ-bit-compression (ℓ-B-Comp) gates, where $3 \le \ell \le k+1$.

3.5   The All-Bonacci Algorithm

In Example 11 ($\mathcal{F}_k$ applied to $k+1$ spins), the resulting biases of the computation spins, $2^{k-1}\varepsilon_0, 2^{k-2}\varepsilon_0, \ldots, 2\varepsilon_0, \varepsilon_0$, were proportional to the exponential series $\{2_{k-1}, 2_{k-2}, \ldots, 4, 2, 1\}$. For example, $\mathcal{F}_2$ on 3 spins results in $\{2\varepsilon_0, \varepsilon_0, \varepsilon_0\}$ (see Example 6), and $\mathcal{F}_3$ on 4 spins results in $\{4\varepsilon_0, 2\varepsilon_0, \varepsilon_0, \varepsilon_0\}$ (see Example 8). This coincides with the k-step Fibonacci sequence, where $a_\ell = 2^{\ell-2}$, for $a_1 = 1$ and $\ell = 2, 3, ..., k+1$ (see eq 28). This property leads to cooling of $n+1$ spins by a special case of k-bonacci (Algorithm 5), where k-bonacci is applied to $k = n-1$ spins.

Algorithm 5: *All-bonacci:* $\mathcal{F}_{All}(A_n, \ldots, A_1, m_n \ldots, m_3)$
Apply $\mathcal{F}_{n-1}(A_n, \ldots, A_1, m_n, \ldots, m_3)$.

The final biases after all-bonacci are $\varepsilon_{A_i} \to \varepsilon_0 2^{i-2}$ for $i > 1$. The resulting series is $\{\ldots, 16, 8, 4, 2, 1, 1\}$.

The all-bonacci algorithm potentially constitutes an optimal AC scheme, as explained in section 4.

3.6   Density Matrices of the Cooled Spin Systems

For a spin system in a completely mixed state, the density matrix of each spin is:

$$\frac{1}{2}\mathcal{I} = \frac{1}{2}\begin{pmatrix} 1 \\ & 1 \end{pmatrix}. \tag{29}$$

The density matrix of the entire system, which is diagonal, is given by the tensor product $\rho_{CM} = \frac{1}{2^3}\mathcal{I} \otimes \mathcal{I} \otimes \mathcal{I}$, where

$$Diag(\rho_{CM}) = 2^{-3}(1, 1, 1, 1, 1, 1, 1, 1). \tag{30}$$

For three spins in a thermal state, the density matrix of each spin is

$$\rho_T^{(1)} = \frac{1}{2}\begin{pmatrix} 1+\varepsilon_0 \\ & 1-\varepsilon_0 \end{pmatrix}. \tag{31}$$

and the density matrix of the entire system is given by the tensor product $\rho_T = \rho_T^{(1)} \otimes \rho_T^{(1)} \otimes \rho_T^{(1)}$. This matrix is also diagonal. We write the diagonal elements to leading order in ε_0:

$$Diag(\rho_T) = 2^{-3}(1 + 3\varepsilon_0, 1 + \varepsilon_0, 1 + \varepsilon_0, 1 + \varepsilon_0, 1 - \varepsilon_0, 1 - \varepsilon_0, 1 - \varepsilon_0, 1 - 3\varepsilon_0). \quad (32)$$

Consider now the density matrix after shifting and scaling, $\rho' = 2^n(\rho - 2^{-n}\mathcal{I})/\varepsilon_0$. For any $Diag(\rho) = (p_1, p_2, p_3, \ldots, p_n)$ the resulting diagonal is $Diag(\rho') = (p'_1, p'_2, p'_3, \ldots, p'_n)$ with $p'_j = 2^n(p_j - 2^{-n})/\varepsilon_0$. The diagonal of a shifted and scaled (S&S) matrix for a completely mixed state (eq 30) is

$$Diag(\rho'_{CM}) = (0, 0, 0, 0, 0, 0, 0, 0), \quad (33)$$

and for a thermal state (eq. 32)

$$Diag(\rho'_T) = (3, 1, 1, 1, -1, -1, -1, -3). \quad (34)$$

In the following discussion we assume that any element, p, satisfies $p'\varepsilon_0 \ll 1$. When applied to a diagonal matrix with elements of the form $p = \frac{1}{2^n}(1 \pm p'_i \varepsilon_0)$, this transformation yields the corresponding S&S elements, $\pm p'i$.

Consider now the application of the Fibonacci to three spins. The resultant bias configuration, $\{2\varepsilon_0, \varepsilon_0, \varepsilon_0\}$ (eq 12) is associated with the density matrix

$$\rho_{Fib}^{(3)} = \frac{1}{2^3}\begin{pmatrix} 1 + 2\varepsilon_0 & \\ & 1 - 2\varepsilon_0 \end{pmatrix} \otimes \begin{pmatrix} 1 + \varepsilon_0 & \\ & 1 - \varepsilon_0 \end{pmatrix} \otimes \begin{pmatrix} 1 + \varepsilon_0 & \\ & 1 - \varepsilon_0 \end{pmatrix}, \quad (35)$$

with a corresponding diagonal, to leading order in ε_0,

$$Diag\left(\rho_{Fib}^{(3)}\right) = 2^{-3}\left(1 + 4\varepsilon_0, 1 + 2\varepsilon_0, 1 + 2\varepsilon_0, 1, 1, 1 - 2\varepsilon_0, 1 - 2\varepsilon_0, 1 - 4\varepsilon_0\right). \quad (36)$$

The S&S form of this diagonal is

$$Diag\left(\rho'^{(3)}_{Fib}\right) = (4, 2, 2, 0, 0, -2, -2, -4). \quad (37)$$

Similarly, the S&S diagonal for Tribonacci on four spins is

$$Diag\left(\rho'^{(4)}_{Trib}\right) = (8, 6, 6, 4, 4, 2, 2, 0, 0, -2, -2, -4, -4, -6, -6, -8). \quad (38)$$

and the S&S form of the diagonal for all-bonacci on n spins is

$$Diag\left(\rho'^{(n)}_{allb}\right) = \left[2^{n-1}, (2^{n-1} - 2), (2^{n-1} - 2), \ldots, 2, 2, 0, 0, \ldots, -2^n - 1\right]. \quad (39)$$

which are good approximations as long as $2^n\varepsilon_0 \ll 1$.

Partner Pairing Algorithm. Recently a cooling algorithm was devised that achieves a superior bias than previous AC algorithms. [18,19] This algorithm, termed the Partner Pairing Algorithm (PPA), was shown to produce the highest possible bias for an arbitrary number of spins after any number of reset steps. Let us assume that the reset spin is the least significant bit (the rightmost spin in the tensor-product density matrix). The PPA on n spins is given in Algorithm 6.

Algorithm 6: *Partner Pairing Algorithm (PPA)*
Repeat the following, until cooling arbitrarily close to the limit.

1. *RESET – applied only to a single reset spin.*
2. *SORT – A permutation that sorts the 2^n diagonal elements of the density matrix by decreasing value, such that p_0 is the largest, and p_{2^n-1} is the smallest.*

Written in terms of ε, the reset step has the effect of changing the traced density matrix of the reset spin to

$$\rho_\varepsilon = \frac{1}{e^\varepsilon + e^{-\varepsilon}} \begin{pmatrix} e^\varepsilon & \\ & e^{-\varepsilon} \end{pmatrix} = \frac{1}{2} \begin{pmatrix} 1+\varepsilon_0 & \\ & 1-\varepsilon_0 \end{pmatrix}, \tag{40}$$

for any previous state. From eq 40 it is clear that $\varepsilon_0 = \tanh \varepsilon$ as stated in the introduction. For a single spin in any diagonal mixed state:

$$\begin{pmatrix} p_0 & \\ & p_1 \end{pmatrix} \xrightarrow{RESET} \frac{p_0 + p_1}{2} \begin{pmatrix} 1+\varepsilon_0 & \\ & 1-\varepsilon_0 \end{pmatrix} = \frac{1}{2} \begin{pmatrix} 1+\varepsilon_0 & \\ & 1-\varepsilon_0 \end{pmatrix}. \tag{41}$$

For two spins in any diagonal state a reset of the least significant bit results in

$$\begin{pmatrix} p_0 & & & \\ & p_1 & & \\ & & p_2 & \\ & & & p_3 \end{pmatrix} \xrightarrow{RESET}$$

$$\frac{p_0 + p_1}{2} \begin{pmatrix} \frac{p_0+p_1}{2}(1+\varepsilon_0) & & & \\ & \frac{p_0+p_1}{2}(1-\varepsilon_0) & & \\ & & \frac{p_2+p_3}{2}(1+\varepsilon_0) & \\ & & & \frac{p_2+p_3}{2}(1-\varepsilon_0) \end{pmatrix}. \tag{42}$$

Algorithm 6 may be highly inefficient in terms of logic gates. Each SORT could require an exponential number of gates. Furthermore, even calculation of the required gates might be exponentially hard.

We refer only to diagonal matrices and the diagonal elements of the matrices (applications of the gates considered here to a diagonal density matrix do not produce off-diagonal elements). For a many-spin system, RESET of the reset spin, transforms the diagonal of any diagonal density matrix, ρ, as follows:

$$Diag(\rho) = (p_0, p_1, p_2, p_3, \ldots) \to \tag{43}$$
$$\left[\frac{p_0 + p_1}{2}(1+\varepsilon_0), \frac{p_0 + p_1}{2}(1-\varepsilon_0), \frac{p_2 + p_3}{2}(1+\varepsilon_0), \frac{p_2 + p_3}{2}(1-\varepsilon_0), \ldots \right],$$

as the density matrix of each pair, p_i and p_{i+1} (for even i) is transformed by the RESET step as described by eq 40 above. We use the definition of S&S probabilities, $p' = 2^n(p - 2-n)/\varepsilon_0$. The resulting S&S diagonal is

$Diag(\rho') =$

$$\left[\frac{2^n}{\varepsilon_0} \left(\frac{p_0 + p_1}{2} + (1 + \varepsilon_0) - 2^{-n} \right), \ldots \right] =$$

$$\left[\frac{2^n}{\varepsilon_0} \left(\frac{p_0 + p_1}{2} - 2^{-n} \right) + 2^n \frac{p_0 + p_1}{2}, \ldots \right] =$$

$$\left[\frac{p_1' + p_2'}{2} + 1, \frac{p_0' + p_1'}{2} + 2^n \frac{p_0 + p_1}{2}, \frac{p_0' + p_1'}{2} - 2^n \frac{p_0 + p_1}{2}, \ldots \right], \quad (44)$$

where the second element is shown in the final expression. We now use $p = 2^{-n}(\varepsilon_0 p' + 1)$, to obtain

$$2^n \frac{p_0 + p_1}{2} = 2^n \frac{2^{-n}(\varepsilon_0 p_0' + 1) + 2^{-n}(\varepsilon_0 p_1' + 1)}{2} \quad (45)$$

$$= \frac{\varepsilon_0(p_0' + p_1') + 2}{2} = 1 + \varepsilon_0 \frac{p_0' + p_1'}{2}. \quad (46)$$

Ref [18] provides an analysis of the PPA. We continue, as in the previous subsection, to analyze the case of $p'\varepsilon_0 \ll 1$, which is of practical interest. In this case $1 + \varepsilon_0 \frac{p_0' + p_1'}{2} \approx 1$. Hence, the effect of a RESET step on the S&S diagonal is:

$$Diag(\rho') = (p_0', p_1', p_2', p_3', \ldots) \rightarrow \quad (47)$$

$$\left[\frac{p_0' + p_1'}{2} + 1, \frac{p_0' + p_1'}{2} - 1, \frac{p_2' + p_3'}{2} + 1, \frac{p_2' + p_3'}{2} - 1, \ldots \right]$$

Consider now three spins, such that the one at the right is a reset spin. Following ref [18], we initially apply the PPA to three spins which are initially at the completely mixed state. The first step of the PPA, RESET (eq 47), is applied to the diagonal of the completely mixed state (eq 33), to yield

$$Diag(\rho'_{CMS}) \rightarrow Diag(\rho'_{RESET}) = (1, -1, 1, -1, 1, -1, 1, -1). \quad (48)$$

This diagonal corresponds to the density matrix

$$\rho_{RESET} = \frac{1}{2^3} \mathcal{I} \otimes \mathcal{I} \otimes \begin{pmatrix} 1 + \varepsilon_0 & \\ & 1 - \varepsilon_0 \end{pmatrix}, \quad (49)$$

namely to the bias configuration $\{0, 0, \varepsilon_0\}$. The next PPA step, SORT, sorts the diagonal elements in decreasing order:

$$Diag(\rho'_{SORT}) = (1, 1, 1, 1, -1, -1, -1, -1), \quad (50)$$

that arises from the density matrix

$$\rho_{SORT} = \frac{1}{2^3} \begin{pmatrix} 1 + \varepsilon_0 & \\ & 1 - \varepsilon_0 \end{pmatrix} \otimes \mathcal{I} \otimes \mathcal{I}, \quad (51)$$

which corresponds to the biases $\{\varepsilon_0, 0, 0\}$. The bias was thus transferred to the leftmost spin. In the course of repeated alternation between the two steps of the PPA, this diagonal will further evolve as detailed in Example 12.

The rightmost column of Example 12 lists the resulting bias configurations. Notice that after the 5th step the biases are identical. Also notice that in the 6th and 10th steps the states $|100\rangle$ and $|011\rangle$ are exchanged (3B-Comp); after both these steps, the state of the system cannot be written as a tensor product.[6] Also notice that after step 13, the PPA applies a SORT which will also switch between these two states. The PPA, applied to the bias configuration $\{t\varepsilon_0, \varepsilon_0, \varepsilon_0\}$, where $1 \leq t \leq 2$ is simply an exchange $|100\rangle \leftrightarrow |011\rangle$. This is evident from the diagonal

$$Diag(\rho') = (t+2, t, t, t-2, -t+2, -t, -t, -t-2). \tag{52}$$

Thus, the PPA on three spins is identical to the Fibonacci algorithm applied to three spins (see Example 6). This analogy may be taken further to some extent. When the PPA is applied onto four spins, the outcome, but not the steps, is identical to the result obtained by Tribonacci algorithm (see Example 8). These identical results may be generalized to the PPA and all-bonacci applied onto $n \geq 3$ spins.

Example 12: *Application of PPA to 3 spins*

		$Diag(\rho') = (0,0,0,0,0,0,0,0)$	$\{0,0,0\}$
step 1	$\xrightarrow{RESET}$	$(1,-1,1,-1,1,-1,1,-1)$	$\{0,0,\varepsilon_0\}$
step 2	$\xrightarrow{SORT}$	$(1,1,1,1,-1,-1,-1,-1)$	$\{\varepsilon_0,0,0\}$
step 3	$\xrightarrow{RESET}$	$(2,0,2,0,0,-2,0,-2)$	$\{\varepsilon_0,0,\varepsilon_0\}$
step 4	$\xrightarrow{SORT}$	$(2,2,0,0,0,0,-2,-2)$	$\{\varepsilon_0,\varepsilon_0,0\}$
step 5	$\xrightarrow{RESET}$	$(3,1,1,-1,1,-1,-1,-3)$	$\{\varepsilon_0,\varepsilon_0,\varepsilon_0\}$
step 6	$\xrightarrow{SORT}$	$(3,1,1,1,-1,-1,-1,-3)$	$N.\,A.$
step 7	$\xrightarrow{RESET}$	$(3,1,2,0,0,-2,-1,-3)$	$\{\frac{3\varepsilon_0}{2}, \frac{\varepsilon_0}{2}, \varepsilon_0\}$
step 8	$\xrightarrow{SORT}$	$(3,2,1,0,0,-1,-2,-3)$	$\{\frac{3\varepsilon_0}{2}, \varepsilon_0, \frac{\varepsilon_0}{2}\}$
step 9	$\xrightarrow{RESET}$	$\frac{1}{2}(7,3,3,-1,1,-3,-3,-7)$	$\{\frac{3\varepsilon_0}{2}, \varepsilon_0, \varepsilon_0\}$
step 10	$\xrightarrow{SORT}$	$\frac{1}{2}(7,3,3,1,-1,-3,-3,-7)$	$N.A.$
step 11	$\xrightarrow{RESET}$	$\frac{1}{2}(7,3,4,0,0,-4,-3,-7)$	$\{\frac{7\varepsilon_0}{4}, \frac{3\varepsilon_0}{4}, \varepsilon_0\}$
step 12	$\xrightarrow{SORT}$	$\frac{1}{2}(7,4,3,0,0,-3,-4,-7)$	$\{\frac{7\varepsilon_0}{4}, \varepsilon_0, \frac{3\varepsilon_0}{4}\}$
step 13	$\xrightarrow{RESET}$	$\frac{1}{4}(15,7,7,-1,1,-7,-7,-15)$	$\{\frac{7\varepsilon_0}{4}, \varepsilon_0, \varepsilon_0\}$

4   Optimal Algorithmic Cooling

4.1   Lower Limits of the PPA and All-Bonacci

Consider an application of the PPA to a three-spin system at a state with the S&S diagonal of eq 37. This diagonal is both sorted and invariant to RESET,

---

[6] This is due to *classical* correlations (not involving entanglement) between the spins; an individual bias may still be associated with each spin by tracing out the others, but such a bias cannot be interpreted as temperature.

hence it is invariant to the PPA. Eq 37 corresponds to the bias configuration $\{2\varepsilon_0, \varepsilon_0, \varepsilon_0\}$ which is the limit of the Fibonacci algorithm presented above (eq 12). This configuration is the "lower" limit of the PPA with three spins in the following sense: any "hotter configuration" is not invariant and continues to cool down during exhaustive PPA until reaching or bypassing this configuration. For four spins the diagonal of Eq 38 is invariant to the PPA for the same reasons. This diagonal corresponds to the bias configuration $\{4\varepsilon_0, 2\varepsilon_0, \varepsilon_0, \varepsilon_0\}$, which is the limit of the Tribonacci algorithm (or the all-bonacci) applied to four spins. Any "hotter configuration" is not invariant and cools further during exhaustive PPA, until it reaches (or bypasses) this configuration.

Now, we follow the approximation of very small biases for n spins, where $2^n \varepsilon_0 \ll 1$. The diagonal in eq 39 is invariant to the PPA. This diagonal corresponds to the bias configuration

$$\{2^{n-2}\varepsilon_0, 2^{n-3}\varepsilon_0, 2^{n-4}\varepsilon_0, \ldots, 2\varepsilon_0, \varepsilon_0, \varepsilon_0\}, \tag{53}$$

which is the limit of the all-bonacci algorithm. As before, any "hotter configuration" is not invariant and cools further. It is thus proven that the PPA reaches at least the same biases as all-bonacci.

We conclude that under the assumption $2^n \varepsilon_0 \ll 1$, the n-spin system can be cooled to the bias configuration shown in eq 53. When this assumption is not valid (e.g., for larger n with the same ε_0), pure qubits can be extracted; see theorem 2 in [19] (theorem 3 in ref [18]).

Yet, the PPA potentially yields "colder configurations". We obtained numerical results for small spin systems which indicate that the limits of the PPA and all-bonacci are identical. Still, since the PPA may provide better cooling, it is important to put an "upper" limit on its cooling capacity.

4.2 Upper Limits on Algorithmic Cooling

Theoretical limits for cooling with algorithmic cooling devices have recently been established [18,19]. For any number of reset steps, the PPA has been shown to be optimal in terms of entropy extraction (see ref [19] and more details in section 1 of ref [18]). An upper bound on the degree of cooling attainable by the PPA is therefore also an upper bound of any AC algorithm. The following theorem regards a bound on AC which is the loose bound of ref [18].

Theorem 1. *No algorithmic cooling method can increase the probability of any basis state to above[7] $\min\{2^{-n}e^{2^n \varepsilon}, 1\}$, where the initial configuration is the completely mixed state.[8] This includes the idealization where an unbounded number of reset and logic steps can be applied without error or decoherence.*

The proof of Theorem 1 involves applying the PPA and showing that the probability of any state never exceeds $2^{-n}e^{2^n \varepsilon}$.

---

[7] A tighter bound, $p_0 \leq 2^{-n}e^{2^{n-1}\varepsilon}$, was claimed by theorem 1 of ref [18].

[8] It is assumed that the computation spins are initialized by the polarization of the reset spins.

5   Algorithmic Cooling and NMR Quantum Computing

We have been using the language of classical bits, however spins are quantum systems; thus, spin particles (two-level systems) should be regarded as quantum bits (qubits). A molecule with n spin nuclei can represent an n-qubit computing device. The quantum computing device in this case is actually an ensemble of many such molecules. In ensemble NMR quantum computing [20,21,22] each computer is represented by a single molecule, such as the TCE molecule of Fig. 1, and the qubits of the computer are represented by nuclear spins. The macroscopic number of identical molecules available in a bulk system is equivalent to many processing units which perform the same computation in parallel. The molecular ensemble is placed in a constant magnetic field, so that a small majority of the spins are aligned with the direction of the field. To perform a desired computation, the same sequence of external pulses is applied to all molecules/computers. Any quantum logic-gate can be implemented in NMR by a sequence of radio-frequency pulses and intermittent delay periods during which spins evolve under coupling [23]. Finally, the state of a particular qubit is measured by summing over all computers/molecules. The process of AC constitutes a simple quantum computing algorithm. However, unlike other quantum algorithms, the use of quantum logic gates does not produce any computational speed-up, but instead generates colder spins. This constitutes the first near-future application of quantum computing devices. AC may also have an important long-term application; it may enable quantum computing devices capable of running important quantum algorithms, such as the factorization of large numbers [24]. NMR quantum computers [20,21,22] are currently the most successful quantum computing devices (see for instance ref [25]), but are known to suffer from severe scalability problems [1,26,27]. AC can be used for building scalable NMR quantum computers of 20-50 quantum bits if electron spins are used for the PT and RESET steps. PT with electron spins [9,11] can enhance the polarization by three or four orders of magnitude. Unfortunately, severe technical difficulties have thus far impeded common practice of this technique. An example of such a difficulty is the need to master two very different electromagnetic frequencies within a single machine. However, in case such PT steps come into practice (using machinery that allows conventional NMR techniques as well), AC could be applied with much better parameters; First, ε_0 could be increased to around 0.01-0.1. Second, the ratio $R_{relax-times}$ could reach $10^3 - 10^4$. With these figures, scalable quantum computers of 20-50 qubits may become feasible.

6   Discussion

Algorithmic Cooling (AC) harnesses the environment to enhance spin polarization much beyond the limits of reversible polarization compression (RPC). Both cooling methods may be regarded as a set of logic gates, such as NOT or SWAP, which are applied onto the spins. Polarization transfer (PT), for instance, a form of RPC, may be obtained by a SWAP gate. AC algorithms are composed of two types of steps: reversible AC steps (RPC) applied to two spins or more, and reset steps, in which entropy is shifted to the environment through reset spins. These reset spins thermalize much faster than computation spins (which are cooled), allowing a form of a molecular heat pump. A prerequisite for AC is thus the mutual existence (on the same molecule) of two types of spins with a substantial relaxation times ratio (of at least an order of magnitude). While this demand limits the applicability of AC, it is often met in practice (e.g., by ^{1}H vs ^{13}C in certain organic molecules) and may be induced by the addition of a paramagnetic reagent. The attractive possibility of using rapidly-thermalizing electron spins as reset bits is gradually becoming a relevant option for the far future.

We have surveyed previous cooling algorithms and suggested novel algorithms for exhaustive AC: the Tribonacci, the k-bonacci, and the all-bonacci algorithms. We conjectured the optimality of all-bonacci, as it appears to yield the same cooling level as the PPA of refs [18,19]. Improving the SNR of NMR by AC potentially constitutes the first short-term application of quantum computing devices. AC is further accommodated in quantum computing schemes (NMR-based or others) relating to the more distant future [1,28,29,30].

Acknowledgements

Y.E., T.M., and Y.W. thank the Israeli Ministry of Defense, the Promotion of Research at the Technion, and the Institute for Future Defense Research for supporting this research.

References

1. Boykin, P.O., Mor, T., Roychowdhury, V., Vatan, F., Vrijen, R.: Algorithmic cooling and scalable NMR quantum computers. Proc. Natl. Acad. Sci. 99(6), 3388–3393 (2002)
2. Fernandez, J.M., Lloyd, S., Mor, T., Rowchoudury, V.: Algorithmic cooling of spins: A practicable method for increasing polarisation. Int. J. Quant. Inf. 2(4), 461–467 (2004)
3. Mor, T., Roychowdhury, V., Lloyd, S., Fernandez, J.M., Weinstein, Y.: US patent No. 6,873,154 (2005)
4. Cover, T.M., Thomas, J.A.: Elements of Information Theory. Wiley, New York (1991)
5. Brassard, G., Elias, Y., Fernandez, J.M., Gilboa, H., Jones, J.A., Mor, T., Weinstein, Y., Xiao, L.: Experimental heat-bath cooling of spins. Proc. Natl. Acad. Sci. USA (submitted) (also in arXiv:quant-ph/0511156)
6. Morris, G.A., Freeman, R.: Enhancement of nuclear magnetic resonance signals by polarization transfer. J. Am. Chem. Soc. 101, 760–762 (1979)
7. Sørensen, O.W.: Polarization transfer experiments in high-resolution NMR spectroscopy. Prog. Nucl. Mag. Res. Spec. 21, 503–569 (1989)
8. Schulman, L.J., Vazirani, U.V.: Scalable NMR quantum computation. In: ACM Symposium on the Theory of Computing (STOC): Proceedings, pp. 322–329. ACM Press, New York (1999)
9. Farrar, C.T., Hall, D.A., Gerfen, G.J., Inati, S.J., Griffin, R.G.: Mechanism of dynamic nuclear polarization in high magnetic fields. J. Chem. Phys. 114, 4922–4933 (2001)

10. Slichter, C.P.: Principles of Magnetic Resonance, 3rd edn. Springer, Heidelberg (1990)
11. Ardenkjær-Larsen, J.H., Fridlund, B., Gram, A., Hansson, G., Hansson, L., Lerche, M.H., Servin, R., Thaning, M., Golman, K.: Increase in signal-to-noise ratio of > 10,000 times in liquid-state NMR. Proc. Natl. Acad. Sci. 100, 10158–10163 (2003)
12. Anwar, M., Blazina, D., Carteret, H., Duckett, S.B., Halstead, T., Jones, J.A., Kozak, C., Taylor, R.: Preparing high purity initial states for nuclear magnetic resonance quantum computing. Phys. Rev. Lett. 93 (2004) (also in arXiv:quant-ph/0312014)
13. Oros, A.M., Shah, N.J.: Hyperpolarized xenon in NMR and MRI. Phys. Med. Biol. 49, R105–R153 (2004)
14. Emsley, L., Pines, A.: Lectures on pulsed NMR. In: Nuclear Magnetic Double Resonance, Proceedings of the CXXIII School of Physics Enrico Fermi, 2nd edn. p. 216. World Scientific, Amsterdam (1993)
15. Elias, Y., Fernandez, J.M., Mor, T., Weinstein, Y.: Algorithmic cooling of spins. Isr. J. Chem. (to be published on 2007)
16. Fernandez, J.M.: De computatione quantica. PhD thesis, University of Montreal, Canada (2003)
17. Weinstein, Y.: Quantum computation and algorithmic cooling by nuclear magnetic resonance. Master's thesis, Physics Department, Technion - Israel Institute of Technology (August 2003)
18. Schulman, L.J., Mor, T., Weinstein, Y.: Physical limits of heat-bath algorithmic cooling. SIAM J. Comp. 36, 1729–1747 (2007)
19. Schulman, L.J., Mor, T., Weinstein, Y.: Physical limits of heat-bath algorithmic cooling. Phys. Rev. Lett. 94, 120501 (2005)
20. Cory, D.G., Fahmy, A.F., Havel, T.F.: Nuclear magnetic resonance spectroscopy: an experimentally accessible paradigm for quantum computing. In: Proceedings of PhysComp96, pp. 87–91 (1996)
21. Cory, D.G., Fahmy, A.F., Havel, T.F.: Ensemble quantum computing by nuclear magnetic resonance spectroscopy. Proc. Natl. Acad. Sci. 1634–1639 (1997)
22. Gershenfeld, N.A., Chuang, I.L.: Bulk spin-resonance quantum computation. Science 275, 350–356 (1997)
23. Price, M.D., Havel, T.F., Cory, D.G.: Multiqubit logic gates in NMR quantum computing. New Journal of Physics 2, 10.1–10.9 (2000)
24. Shor, P.W.: Polynomial-time algorithms for prime factorization and discrete logarithms on a quantum computer. SIAM J. Comp. 26(5), 1484–1509 (1997)
25. Vandersypen, L.M.K., Steffen, M., Breyta, G., Yannoni, C.S., Sherwood, M.H., Chuang, I.L.: Experimental realization of Shor's quantum factoring algorithm using nuclear magnetic resonance. Nature 414, 883–887 (2001)
26. Warren, W.S.: The usefulness of NMR quantum computing. Science 277, 1688–1690 (1997)
27. DiVincenzo, D.P.: Real and realistic quantum computers. Nature 393, 113–114 (1998)
28. Twamley, J.: Quantum-cellular-automaton quantum computing with endohedal fullerenes. Phys. Rev. A 67, 052318 (2003)
29. Freegarde, T., Segal, D.: Algorithmic cooling in a momentum state quantum computer. Phys. Rev. Lett. 91, 037904 (2003)
30. Ladd, T.D., Goldman, J.R., Yamaguchi, F., Yamamoto, Y., Abe, E., Itoh, K.M.: All-silicon quantum computer. Phys. Rev. Lett. 89, 017901 (2002)

Nanocomputing by Self-assembly

Lila Kari

Department of Computer Science
University of Western Ontario
London, Ontario
Canada N6A 5B7
lila@csd.uwo.ca

Abstract. Biomolecular (DNA) computing is an emergent field of un-conventional computing, lying at the crossroads of mathematics, computer science and molecular biology. The main idea behind biomolecular computing is that data can be encoded in DNA strands, and techniques from molecular biology can be used to perform arithmetic and logic operations. The birth of this field was the 1994 breakthrough experiment of Len Adleman who solved a hard computational problem solely by manipulating DNA strands in test-tubes. This led to the possibility of envisaging a DNA computer that could be thousand to a million times faster, trillions times smaller and thousand times more energy efficient than today's electronic computers.

I will present one of the most active directions of research in DNA computing, namely DNA nanocomputing by self-assembly. I will namely discuss the computational potential of self-assembly, the process by which objects autonomously come together to form complex structures. I will bring forth evidence that self-assembly of DNA molecules can be used to perform computational tasks. Lastly, I will address the problem of self-assembly of arbitrarily large super-shapes, its solution and implications.

S.G. Akl et al.(Eds.): UC 2007, LNCS 4618, p. 27, 2007.
© Springer-Verlag Berlin Heidelberg 2007

Organic User Interfaces (Oui!): Designing Computers in Any Way Shape or Form

Roel Vertegaal

Xuuk Inc. / Human Media Laboratory
School of Computing, Queen's University
Kingston, Ontario
Canada K7L 3N6
roel@xuuk.com

Abstract. Over the past few years, there has been a quiet revolution in display manufacturing technology. One that is only comparable in scope to that of the invention of the first LCD, which led to DynaBook and the modern laptop. E-ink electrophoretic pixel technology, combined with advances in organic thin-film circuit substrates, have led to displays that are so thin and flexible they are beginning to resemble paper. Soon displays will completely mimic the high contrast, low power consumption and flexibility of printed media. As with the invention of the first LCD, this means we are on the brink of a new paradigm in computer user interface design: one in which computers can have any organic form or shape. One where any object, no matter how complex, dynamic or flexible its structure, may display information. One where the deformation of shape is a main source of input.

This new paradigm of Organic User Interface (Oui!) requires a new set of design guidelines, which I will discuss in this presentation. These guidelines were inspired by architecture, which went through a similar transformation decades ago. In Oui! The Input Device Is The Output Device (TIDISTOD), Form dynamically follows Flow of activities of the human body, and Function equals Form. I will give an overview of technologies that led to Organic UI, such as Tangible UI and Digital Desks, after which I will discuss some of the first real Oui! interfaces, which include Gummi and PaperWindows. PaperWindows, which was developed at HML, is the first real paper computer. It uses computer vision to track sheets of real paper in real time. Page shape is modeled in 3D, textured with windows and projected back onto the paper, making for a wireless hi-res flexible color display. Interactions with PaperWindows take place through hand gestures and paper folding techniques.

S.G. Akl et al.(Eds.): UC 2007, LNCS 4618, p. 28, 2007.
© Springer-Verlag Berlin Heidelberg 2007

Unconventional Models of Computation
Through Non-standard Logic Circuits

Juan C. Agudelo[1] and Walter Carnielli[2]

[1] Ph.D. Program in Philosophy/Logic
IFCH and Group for Applied and Theoretical Logic- CLE
State University of Campinas - UNICAMP, Brazil
[2] IFCH and Group for Applied and Theoretical Logic- CLE
State University of Campinas - UNICAMP, Brazil
SQIG - IT, Portugal
{juancarlos,carniell}@cle.unicamp.br

Abstract. The classical (boolean) circuit model of computation is generalized via polynomial ring calculus, an algebraic proof method adequate to non-standard logics (namely, to all truth-functional propositional logics and to some non-truth-functional logics). Such generalization allows us to define models of computation based on non-standard logics in a natural way by using 'hidden variables' in the constitution of the model. *Paraconsistent circuits* for the paraconsistent logic *mbC* (and for some extensions) are defined as an example of such models. Some potentialities are explored with respect to computability and computational complexity.

1 Introduction

The classical notion of *algorithm* is founded in the model of automated machines introduced by Turing in [18], now known as Turing machines, and in its equivalent theories (λ-definability, partial recursive functions, uniform families of boolean circuits, and so on). Surprising connections between classical logic and classical computation have been established, as the equivalence between the well-known 'halting problem' and the undecidability of first-order logic (see [18], [3] and [2]), and the relationship between computational complexity and expressibility in logic (see [15, sec. 2] for a survey). The advent of so many non-classical logics challenge us to think, prompted by the above connections, in the possibilities of logic relativization of the notion of computability and in its conceivable advantages. In this spirit we presented in [1] a model of 'paraconsistent Turing machines', a generalization of Turing machines through a paraconsistent logic, and proved that some characteristics of quantum computation can be simulated by this kind of machines. In particular, we have showed how this model can be used to solve the so-called Deutsch-Jozsa problem in an efficient way. In this paper we propose another generalization of a classical model of computation (boolean circuits), defining unconventional models of logic circuits where gate operations are defined in accordance with adequate semantics for non-classical

S.G. Akl et al.(Eds.): UC 2007, LNCS 4618, pp. 29–40, 2007.
© Springer-Verlag Berlin Heidelberg 2007

propositional logics. The conspicuous unconventional peculiarities of our approach are represented by the use of non-truth functional logics to express logic circuitry (instead of a physical or biological approach) and by the significance of such logical standpoint in the foundations of the notion of computability. Some initial inquires about advantages of this logic relativized notion of computability are also addressed here.

A *boolean circuit* is basically a finite collection of input variables and logic gates acyclically connected, where input variables can take values in $\{0, 1\}$ (0 representing the truth value $false$ and 1 representing the truth value $true$), and each gate performs a boolean operation (e.g. AND, OR, NOT). Boolean circuits can be viewed as computing *boolean functions* $f: \{0, 1\}^n \rightarrow \{0, 1\}$. A particular computation is performed by establishing the values of the input variables and reading the result of the computation as the output at the final gate (the gate with no output connections to any other gate).[1] It is clear that the classical (boolean) circuit model of computation is based on classical propositional logic, considering that gates operate in accordance with functional definitions of the classical logic connectives. The fact that uniform boolean families of circuits are a model for Turing machines clarifies much of the contents of celebrated Cook's theorem in [10]; in an analogous way, the L-circuits introduced in Section 3 could shed light in certain aspects of computation related to non-standard logics.

In order to generalize boolean circuits, it is necessary to specify the input-output alphabet (i.e. the set of values allowed for input variables and for outputs of gates) as well as the gate operations. An obvious generalization of boolean circuits to truth-functional many-valued logics would consist in considering the set of truth values as the input-output alphabet, and defining logic gate operations as the functional definitions of the corresponding logic operation. In this way, a logic circuit based in a many-valued logic with truth values set A would compute functions of the form $f: A^n \rightarrow A$.

In the case of infinite-valued logics, despite technical difficulties for implementation of infinitely many distinguishable symbols (for inputs and outputs), the above generalization seems to be adequate for every many-valued logics; however, the specification of gate operations is not obvious for logics without truth-functional semantics.

The original logic relativization of boolean circuits that we propose here is obtained via the *polynomial ring calculus* (PRC) introduced in [6], which is an algebraic proof method basically consisting on the translation of logic formulas into polynomials and transforming deductions into polynomial operations. As shown in [6], PRC is a mechanizable proof method particularly apt for all finitely-many-valued logics and for several non-truth-functional logics as well, provided that they can be characterized by two-valued *dyadic semantics* (see [5]). The generalization of boolean circuits we are going to propose takes advantage of the features of PRC, allowing to define logic gate operations through polynomial operations, and restricting input-output values to finite sets. Interesting

---

[1] This definition can be easily extended to compute functions of the form $f: \{0, 1\}^n \rightarrow \{0, 1\}^m$, allowing more than one final gate.

examples of our generalization are presented by exhibiting PRC for the paraconsistent logic mbC and for some extensions, while exploring some potentialities of this model with respect to computability and computational complexity.

We present a summary of the PRC proof method and some illustrative examples following [6], before presenting the promised generalization of boolean circuits.

2   Polynomial Ring Calculus

PRC consists in translating logic formulas into polynomials over the finite (Galois) fields $GF(p^n)$ (where p is a prime number and n is a natural number).[2] PRC defines rules to operate with polynomials. The first group of rules, the *ring rules*, correspond to the ring properties of addition and multiplication: addition is associative and commutative, there is a 'zero' element and all elements have 'addition inverse'; multiplication is associative, and there is a 'one' element and multiplication distributes over addition. The second group of rules, the *polynomial rules*, establishes that the addition of an element x exactly p^n times can be reduced to the constant polynomial 0; and that elements of the form $x^i \cdot x^j$ can be reduced to $x^k (\bmod\ q(x))$, for $k \equiv i + j (\bmod\ (p^n - 1))$ and $q(x)$ a convenient primitive polynomial (i.e. and irreducible polynomial of degree n with coefficients in $\mathbb{Z}_p$). Two inference metarules are also defined, the *uniform substitution*, which allows to substitute some variable in a polynomial by another polynomial (in all occurrences of the variable), and the *Leibniz rule*, which allows to perform substitutions by 'equivalent' polynomials.

The tip to define a PRC, for an specific logic, is to specify translation of formulas into polynomials in such way that they mimic the conditions of an adequate (correct and complete) class of valuations for the logic in question. We will illustrate this point with some examples.

First, let us consider Classical Propositional Logic (CPL). Denoting by For the set of well formed formulas of CPL, and using Greek letters as metavariables for formulas, a CPL valuation is a function $v\colon For \rightarrow \{0,1\}$ subject to the following conditions (considering For over the alphabet $\{\wedge, \vee, \neg\}$):

$$v(\varphi \wedge \psi) = 1 \text{ iff } v(\varphi) = 1 \text{ and } v(\psi) = 1; \qquad (1)$$

$$v(\varphi \vee \psi) = 1 \text{ iff } v(\varphi) = 1 \text{ or } v(\psi) = 1; \qquad (2)$$

$$v(\neg\varphi) = 1 \text{ iff } v(\varphi) = 0. \qquad (3)$$

In this case, For could be translated to polynomials in the polynomial ring $\mathbb{Z}_2[X]$ (i.e. polynomials with coefficients in the field $\mathbb{Z}_2$ and variables in the set $X = \{x_1, x_2, \ldots\}$). The translation function $*\colon For \rightarrow \mathbb{Z}_2[X]$ is defined by:

---

[2] The number p is usually called the *characteristic* of the field. The characteristic of any finite field is necessarily a prime number, see for example [16].

$$p_i^* = x_i \text{ if } p_i \text{ is a propositional variable;} \qquad (4)$$
$$(\varphi \wedge \psi)^* = \varphi^* \cdot \psi^*; \qquad (5)$$
$$(\varphi \vee \psi)^* = \varphi^* \cdot \psi^* + \varphi^* + \psi^*; \qquad (6)$$
$$(\neg\varphi)^* = \varphi^* + 1. \qquad (7)$$

Polynomial rules in this case establish that $x + x$ can be reduced to 0 and that $x \cdot x$ can be reduced to x. Assigning values in $\{0, 1\}$ to variables in X, it can be easily shown that:

$$(\varphi \wedge \psi)^* = 1 \text{ iff } \varphi^* = 1 \text{ and } \psi^* = 1; \qquad (8)$$
$$(\varphi \vee \psi)^* = 1 \text{ iff } \varphi^* = 1 \text{ or } \psi^* = 1; \qquad (9)$$
$$(\neg\varphi)^* = 1 \text{ iff } \varphi^* = 0. \qquad (10)$$

which means that the translation $*$ characterizes all CPL-valuations. Consequently, using $\varphi(p_1, \ldots, p_k)$ to express that φ is a formula with propositional variables in $\{p_1, \ldots, p_k\}$, we have the following theorem:

Theorem 1. $\vdash_{CPL} \varphi(p_1, \ldots, p_k)$ *iff* $\varphi^*(a_1, \ldots, a_k) = 1$ *for every* $(a_1, \ldots, a_k) \in \mathbb{Z}_2^k$.

And by theorem 2.3 in [6], we have:

Theorem 2. $\vdash_{CPL} \varphi$ *iff* φ^* *reduces by* PRC *rules to the constant polynomial* 1.

As an example of a proof of $\varphi \vee \neg\varphi$ in CPL using PRC (taking into account Theorem 2):

$$
\begin{aligned}
(\varphi \vee \neg\varphi)^* &= \varphi^* \cdot (\neg\varphi)^* + \varphi^* + (\neg\varphi)^* \\
&= \varphi^* \cdot (\varphi^* + 1) + \varphi^* + \varphi^* + 1 \\
&= \varphi^* \cdot \varphi^* + \varphi^* + \varphi^* + \varphi^* + 1 \\
&= \varphi^* + \varphi^* + \varphi^* + \varphi^* + 1 \\
&= 1
\end{aligned}
$$

Now, let us consider the paraconsistent logic mbC, a fundamental logic in the hierarchy of Logics of Formal Inconsistency ($LFIs$). Albeit not a finite valued logic, mbC can be characterized by a non-truth-functional two-valued valuation semantics (cf. [7], Sections 3.2 and 3.3). In this case, valuations are subject to the following conditions (considering For over the alphabet $\{\wedge, \vee, \rightarrow, \neg, \circ\}$, where $\circ$ denotes the 'consistency' operator):

$$v(\varphi \wedge \psi) = 1 \text{ iff } v(\varphi) = 1 \text{ and } v(\psi) = 1; \qquad (11)$$
$$v(\varphi \vee \psi) = 1 \text{ iff } v(\varphi) = 1 \text{ or } v(\psi) = 1; \qquad (12)$$
$$v(\varphi \rightarrow \psi) = 1 \text{ iff } v(\varphi) = 0 \text{ or } v(\psi) = 1; \qquad (13)$$
$$v(\neg\varphi) = 0 \text{ implies } v(\varphi) = 1; \qquad (14)$$
$$v(\circ\varphi) = 1 \text{ implies } v(\varphi) = 0 \text{ or } v(\neg\varphi) = 0. \qquad (15)$$

Note that implications in conditions (14) and (15) hold in one direction only, the other direction being 'indeterminate'. The way offered in [6] to mimic such non-determination via polynomials is the introduction of new variables (outside the set of variables for propositional variables) in the translations from formulas into polynomials. In this paper, we will dub such new variables 'hidden variables'.[3] The simple but powerful strategy of introducing hidden variables is the key idea for making PRC apt for a wide class of non-classical logics, including non-truth-functional and infinite-valued logics. Using this strategy, formulas in mbC can be translated into polynomials in the polynomial ring $\mathbb{Z}_2[X]$, defining the translation function $*$: $For \rightarrow \mathbb{Z}_2[X]$ by:

$$p_i^* = x_i \text{ if } p_i \text{ is a propositional variable;} \tag{16}$$

$$(\varphi \wedge \psi)^* = \varphi^* \cdot \psi^*; \tag{17}$$

$$(\varphi \vee \psi)^* = \varphi^* \cdot \psi^* + \varphi^* + \psi^*; \tag{18}$$

$$(\varphi \rightarrow \psi)^* = \varphi^* \cdot \psi^* + \varphi^* + 1; \tag{19}$$

$$(\neg\varphi)^* = \varphi^* \cdot x_\varphi + 1; \tag{20}$$

$$(\circ\varphi)^* = (\varphi^* \cdot (x_\varphi + 1) + 1) \cdot x_{\varphi'}; \tag{21}$$

where x_φ and $x_{\varphi'}$ are hidden variables.[4]

For paraconsistent logics extending mbC where the formula $\circ\varphi$ is equivalent to the formula $\neg(\varphi \wedge \neg\varphi)$ (such as C_1, $mCil$, Cil, $Cile$, $Cila$, $Cilae$ and $Cilo$, see [7]; here we will refer to these logics as mbC^+ logics), the translations of $\circ\varphi$ and $\neg(\varphi \wedge \neg\varphi)$ must be equivalent and dependent of the same variables. For mbC, the translation of $\circ\varphi$ is $(\circ\varphi)^* = (\varphi^* \cdot (x_\varphi + 1) + 1) \cdot x_{\varphi'}$, and the translation of $\neg(\varphi \wedge \neg\varphi)$ is $(\neg(\varphi \wedge \neg\varphi))^* = \varphi^* \cdot (\varphi^* \cdot x_\varphi + 1) \cdot x_{\varphi \wedge \neg\varphi} + 1$. The only variables that do not match in these translations are $x_{\varphi'}$ and $x_{\varphi \wedge \neg\varphi}$. Then, for mbC^+ logics, setting these variables as equal and taking into account that both translations must be equivalent, we have that $x_{\varphi'} = x_{\varphi \wedge \neg\varphi} = 1$, and as consequence $(\circ\varphi)^* = \varphi^* \cdot (x_\varphi + 1) + 1$.

Some valuation semantics introduce conditions over schemes of formulas with more than one logic operator. For instance, paraconsistent logics with the axiom $\neg\neg\varphi \rightarrow \varphi$ must have the following clause in its valuation semantic (cf. [7, p. 58]):

$$v(\neg\neg\varphi) = 1 \text{ implies } v(\varphi) = 1. \tag{22}$$

This kind of condition translate in 'polynomial conditions', i.e. relations between polynomials that must be considered in any polynomial reduction. For mbC, using the PRC presented above, the translation of the formula $\neg\neg\varphi$ is $(\neg\neg\varphi)^* = (\varphi^* \cdot x_\varphi + 1) \cdot x_{\neg\varphi} + 1$. For a logic extending mbC, with a valuation semantic including (22), the following polynomial condition must be taken into account:

---

[3] In [6] such variables are called 'quantum variables', by resemblance with the 'hidden variables' theories of quantum mechanics.

[4] In the translation defined to mbC in [6] a different rule to translate $\circ\varphi$ is presented, but such translation does not permit that $(\circ\varphi)^*$ and φ^* take simultaneously the value 0, while the semantic valuation for mbC permits $v(\circ\varphi) = v(\varphi) = 0$. Our definition fix this problem.

$$(\varphi^* \cdot x_\varphi + 1) \cdot x_{\neg\varphi} = 0 \text{ implies } \varphi^* = 1. \tag{23}$$

3   Generalizing Boolean Circuits Via Polynomial Ring Calculus

Having presented the PRC, it is now easy to generalize boolean circuits to a wide extent of non-classical logics:

Definition 1 (L-circuit). *Let L be a propositional logic provided with a PRC over the finite field F. An L-circuit is a finite directed acyclic graph $C = (V, E)$, where V is the set of nodes and E the set of edges. Nodes without incoming edges are called* inputs *of the circuit, and are denoted by variables (x_1, x_2, ...) or by constants (elements of F). Other nodes are called* logic gates, *and correspond to logic operators of L. A logic gate evaluates the polynomial associated to the corresponding logic operator and perform polynomial reductions in accordance with the PRC for L. The gate without output connections to any other gate gives the output of the circuit.*

It is to be noted that any logic gate has at most one outcoming edge, and that the number of incoming edges of a logic gate is determined by the arity of the corresponding logic operator.

With this definition of L-circuit and the PRC presented to CPL in the previous section, it is easy to see that CPL-circuits behave just in the same way as boolean circuits, as it would have to be expected. More interesting cases of L-circuits are obtained when we consider the paraconsistent logic mbC and its extensions, and their respective PRCs. For instance, the mbC-circuit for the formula $\neg p_1 \wedge \neg p_2$ (graphically represented by the Figure 1) shows how 'hidden variables' appear in the process of computation, giving place to 'indeterminism', an interesting characteristic not present in boolean circuits.[5]

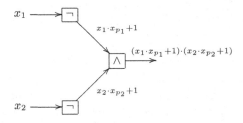

Fig. 1. mbC-circuit for the formula $\neg p_1 \wedge \neg p_2$

---

[5] Probabilistic circuits are defined by introducing probabilistic input variables, but such variables are independent of the logic operations (see [9, Sect. 2.12]). In other words, indeterminism in probabilistic circuits is introduced by allowing indeterminism in the inputs, but not generated by logic operations as in our L-circuits.

This mbC-circuit, when variables x_1 and x_2 take both the value 0, produces the output 1 in a deterministic way. But when x_1 takes the value 1, the output depends on the hidden variable x_{p_1}; and when x_2 takes the value 1, the output depends on the hidden variable x_{p_2}. Hidden variables, like input variables, have to take values in F (the field used in the PRC, in this case $\mathbb{Z}_2$), and we could assume that values to hidden variables are randomly assigned. Then, we could distinguish between *deterministic* and *non-deterministic* L-circuits:

Definition 2 (Deterministic and non-deterministic L-circuit). *An L-circuit is* deterministic *if the reduced polynomial of the output gate does not contain hidden variables, otherwise the L-circuit is* non-deterministic.

Note that there may be deterministic L-circuits with hidden variables within the circuit. For example, the mbC-circuit for the formula $p_1 \vee \neg p_1$, graphically represented by the following figure:

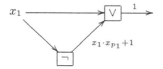

Fig. 2. mbC-circuit for the formula $p_1 \vee \neg p_1$

It is also important to take into account that non-deterministic L-circuits could behave deterministically for some specific inputs, like the mbC-circuit for the formula $\neg p_1 \wedge \neg p_2$ presented above.

Now, having defined the concept of L-circuit, it is opportune to discuss its potentialities. Main characteristics of any model of computation concerns its computability power (the class of functions the model of computation can express) and its computational complexity. Classically, it is possible to show that to every boolean function whose inputs have a fixed length, there is a boolean circuit that computes such function (see for example [17, p. 79, prop. 4.3]). Non-computable functions arise when we consider functions whose inputs have arbitrary length. In this case it is necessary to define a *family of circuits*, which is an infinite denumerable set of circuits $\{C_i\}_{i \in \mathbb{N}}$ where the circuit C_i computes inputs of length i. To accept a family of circuits as an effective procedure we have to impose the restriction that such a family of circuits should be *uniformly generated*, i.e. described by a Turing machine. Usually, uniform generation also imposes computational complexity restrictions; it establishes, for example, that the circuit must be generated by a polynomial time Turing machine, i.e. the time taken to generate the circuit C_i must be a polynomial expression on i. A uniformly generated family of circuits is called a *uniform family* of circuits. It can be proven that uniform families of boolean circuits are equivalent (with respect to computability and computational complexity) to Turing machines (see [4]).

The next two sections explore potentialities of uniform families of L-circuits as models of computation, first with respect to computability and then with respect to computational complexity, considering only to the particular case of paraconsistent logics.

3.1 Potentialities with Respect to Computability

Because paraconsistent logics are deductively weaker than CPL (or a version of CPL in a different signature, see [7, Sect. 3.6]), we could think that the class of functions computed by uniform families of paraconsistent L-circuits would be narrower than functions computed by uniform families of boolean circuits, but we will show that this is not the case for mbC.

Theorem 3. *To every function computed by a uniform family of boolean circuits there is a uniform family of mbC-circuits that computes such function.*

Proof. Let the mapping $t_1 \colon For \to For^\circ$ (where For is the set or formulas of CPL and For° is the set or formulas of mbC) defined as:

1. $t_1(p) = p$, for any propositional variable p;
2. $t_1(\varphi \# \psi) = t_1(\varphi) \# t_1(\psi)$, if $\# \in \{\wedge, \vee, \to\}$;
3. $t_1(\neg\varphi) =\sim t_1(\varphi)$, where $\sim$ is defined in mbC as $\sim \varphi = \varphi \to (\psi \wedge (\neg\psi \wedge \circ\psi))$.

t_1 conservatively translates CPL within mbC (cf. [7, p. 47]), that is, for every $\Gamma \cup \{\varphi\} \subseteq For$:

$$\Gamma \vdash_{CPL} \varphi \text{ iff } t_1(\Gamma) \vdash_{mbC} t_1(\varphi),$$

where $t_1(\Gamma) = \{t_1(\psi) : \psi \in \Gamma\}$.

Using t_1, the uniform family of boolean circuits could be algorithmically translated to an equivalent uniform family of mbC-circuits, just by changing the NOT gates for subcircuits corresponding to the formula $\varphi \to (\psi \wedge (\neg\psi \wedge \circ\psi))$. It can be checked that $(\varphi \to (\psi \wedge (\neg\psi \wedge \circ\psi)))^* = \varphi^* + 1$, which is the same polynomial as in classical negation. □

The previous theorem can be generalized to several other paraconsistent logics, that is, to those where a conservatively translation function from CPL can be defined taking into account that such translation function must be effectively calculated. In the other direction, the existence of uniform families of boolean circuits to every uniform family of L-circuits (for any logic L provided with PRC) is guaranteed by the classical computability of roots for polynomials over finite fields. Then, the L-circuits model does not invalidate Church-Turing's thesis.

Generalizing PRC to infinite fields could be a way to define models of 'hyper-computation' (i.e. models that permit the computation of non Turing-machine computable functions, see [11]). For the case where the field is the rational numbers with usual addition and multiplication, the output of a certain subclass of L-circuits would be Diophantine polynomials, and thus the problem of determining when L-circuits have 0 as output will be equivalent to the problem of determining when Diophantine equations have solutions, a well-known classically unsolvable problem. However, in spite of the possibility of finding a clever way (using hidden variables and polynomial operations) to determine 0 outputs of L-circuits, hypercomputation would be attained at the cost of introducing an infinite element into the model, which seems to be opposed to the intuitive idea of 'computable' (but most models of hypercomputation, nevertheless, are endowed with some kind of infinite element–see [12] and [13] for critiques of hypercomputation).

3.2   Potentialities with Respect to Computational Complexity

Computational complexity potentialities can be thought in both directions too, that is: are there uniform families of boolean circuits more efficient than uniform families of paraconsistent L-circuits? And vice versa? The first question has an immediate answer, but the other is not obvious. We will next show the answer to the first question, and subsequently some ideas related to the second.

Theorem 4. *To every function computed by a polynomial size uniform family of boolean circuits there is a polynomial size uniform family of mbC-circuits that computes such function.*

Proof. The translation t_1, defined in the proof of Theorem 3 adds only a polynomial number of gates to the circuit (4 gates for every NOT gate).   □

As in the case of Theorem 3, Theorem 4 can be generalized to several other paraconsistent logics, those to where a conservative translation function from CPL can be defined (taking into account that such translation function must be effectively calculated and add at most a linear number of gates to the circuits).

The other direction is left as an open problem, but we will show some analogies between *quantum circuits* (see [8]) and paraconsistent circuits that could be valuable in looking for a positive answer.

The quantum circuit model of computation is a generalization of the boolean circuits that depart from a quantum mechanics perspective. In a brief description of such model of computation, we can say that classical *bits* are substituted by *qubits* (or *quantum bits*). Qubits, in a similar way than classical bits, can take values $|0\rangle$ and $|1\rangle$ (where $|0\rangle$ and $|1\rangle$ represent bases vectors in a two dimensional Hilbert space), but the radical difference is that qubits admit also *superpositions* (or linear combinations) of such values, following the rules of the description of a quantum system in quantum mechanics. Superposition of values are usually referred to as *superposition states*. In some interpretations of quantum mechanics (as in the many-worlds interpretations), the states involved in a superposition state are interpreted as coexisting, and then we can think that qubits can take simultaneously the values $|0\rangle$ and $|1\rangle$, with probabilities associated to each value. In quantum circuits, boolean gates are substituted by *quantum gates*, which are unitary operators over qubits, also in agreement with quantum mechanics. A relevant quantum gate is the so-called *Hadamard gate*, which transforms base states into perfect superposition states (states where probabilities are the same to any base state). Because quantum gates are also linear operators, the application of a gate to a superposition state is equivalent to the simultaneous application of the gate to every base state in the superposition.

This characteristic allows the circuit to compute in parallel, in what is called *quantum parallelism*. The problem is that when a measurement is performed, only one of the superposition states is obtained with a probability given by the coefficients of the linear combination. Then, quantum algorithms must take advantage (before performing the measurement) of global characteristics of the functions computed in parallel.

We propose that outputs depending on hidden variables in L-circuits correspond to superposition states, since the indeterminism introduced by hidden variables allows us to simulate indeterminism of superposition states measurements. Under this assumption, gates corresponding to $\neg$ and $\circ$ in mbC and mbC^+ logics, setting 1 as input, could be used to simulate the Hadamard gate (because in mbC, $(\neg\varphi)^*(1) = x_\varphi + 1$ and $(\circ\varphi)^*(1) = x_\varphi \cdot x_{\varphi'}$, and in mbC^+ logics we have that $(\circ\varphi)^*(1) = x_\varphi$). To illustrate how $\neg$ and $\circ$ gates could be used in mbC-circuits and mbC^+-circuits, we will demonstrate a relevant theorem:

Theorem 5. *Let φ a formula in mbC and $\varphi[p_i/\psi]$ the formula obtained from φ by uniformly replacing p_i by ψ. Then, $\varphi^* = f(x_1, \ldots, x_i, \ldots, x_n)$ implies that $(\varphi[p_i/\neg p_i])^*(x_1, \ldots, 1, \ldots, x_n) = f(x_1, \ldots, x_{p_i} + 1, \ldots, x_n)$.*

Proof. By induction on the complexity of φ. $\square$

Theorem 5 shows that, for mbC, by uniformly replacing in a formula φ the propositional variable p_i with the formula $\neg p_i$, and setting 1 to the input variable x_i in the corresponding mbC-circuit, we have that the output polynomial of the mbC-circuit is equal to the polynomial for φ replacing the input variable x_i with $x_{p_i} + 1$.

A similar theorem can be established for mbC^+ logics, uniformly replacing the propositional variable p_i with the formula $\circ p_i$. In this case, the output polynomial is equal to the polynomial for φ replacing the input variable x_i with the hidden variable x_{p_i}. This result could be used to compute satisfiability of CPL formulas in a non-deterministic way, taking the translation of a CPL formula to a mbC^+ logic (with a conservative translation), by uniformly replacing all propositional variables p_i with $\circ p_i$ and setting 1 to all input variables.

For example, the translation of the CPL formula $\varphi = p_1 \wedge p_2$ to a mbC^+ formula (using the translation t_1 above) is the same formula $\varphi = p_1 \wedge p_2$. Using the PRC for mbC, and the translation of $\circ\varphi$ to polynomials in mbC^+ logics already present above, we have that $\varphi^* = x_1 \cdot x_2$ and $(\varphi[p_1/\circ p_1, p_2/\circ p_2])^*(1,1) = x_{p_1} \cdot x_{p_2}$. Note that the only thing we did was to replace input variables with hidden variables.

Obviously, for any CPL formula φ with propositional variables in $\{p_1, \ldots, p_n\}$, if φ is satisfiable, then $(t_1(\varphi)[p_1/\circ p_1, \ldots, p_n/\circ p_n])^*(1, \ldots, 1) = 1$ for some assignment of values to hidden variables. An interesting question is whether there exists any method, in accordance with PRC, that allows to eliminate hidden variables in such a way that polynomials distinct to the constant polynomial 0 reduces to the constant polynomial 1. For some paraconsistent logics that we have explored, this seems to be impossible, but we also do not have a definite negative result (the possibility of polynomial conditions, see last paragraph of

Section 2, could be the key to obtain positive or negative results). We also have to take into account that $PRCs$ can be defined over distinct fields, and for other (not necessarily paraconsistent) non-standard logics.

4   Final Comments

This paper proposes a natural way to generalize boolean circuit model of computation to models based on non-standard logics. This opens possibilities to integrate innovative concepts of non-standard logics into computational theory, shedding light into the question of how depending on logic abstract notions of computability and algorithmic complexity could be. Moreover, the generalization proposed here has the advantage of being flexible enough to be almost immediately adapted to a wide range of non-standard logics.

Cook's theorem (cf. [10]) states that any NP-problem can be converted to the satisfiability problem in CPL in polynomial time. The proof shows, in a constructive way, how to translate a Turing machine into a set of CPL formulas in such a way that the machine outputs '1' if, and only if, the formulas are consistent. As mentioned in the introduction, a model of paraconsistent Turing machines presented in [1] was proved to solve Deutsch-Jozsa problem in an efficient way. We conjecture that a similar result as Cook's theorem can be proven to paraconsistent Turing machines. In this way, paraconsistent circuits could be shown to efficiently solve Deutsch-Jozsa problem. Consequences of this approach would be the definition of 'non-standard' complexity classes relative to such unconventional models of computation founded over non-classical logics.

Finally, another relevant question that we do not tackle in this paper refers to physical viability: are there any physical implementations to the hidden-variable model of computation presented here? We think that hidden-variable theories of quantum mechanics (see [14]) could give a positive response to this question.

Acknowledgements

This research was supported by FAPESP- Fundao de Amparo Pesquisa do Estado de So Paulo, Brazil, Thematic Research Project grant 2004/14107-2. The first author was also supported by a FAPESP scholarship grant 05/05123-3, and the second by a CNPq (Brazil) Research Grant 300702/2005-1 and by an EU-FEDER grant via CLC (Portugal).

References

1. Agudelo, J.C., Carnielli, W.: Quantum algorithms, paraconsistent computation and Deutsch's problem. In: Prasad, B. (ed.) Proceedings of the 2nd Indian International Conference on Artificial Intelligence, Pune, India, pp. 1609–1628 (2005)
2. Boolos, G., Jeffrey, R.: Computability and logic, 3rd edn. Cambridge University Press, Cambridge (1989)

3. Richard Büchi, J.: Turing-machines and the Entscheidungsproblem. Mathematische Annalen 148, 201–213 (1962)
4. Calabro, C.: Turing Machine vs. RAM Machine vs. Circuits. Lecture notes, available at http://http://www-cse.ucsd.edu/classes/fa06/cse200/ln2.ps
5. Coniglio Carlos Caleiro, M.E., Carnielli, W., Marcos, J.: Two's company: the humbug of many logical values. In: Beziau, J.-Y. (ed.) Logica Universalis, pp. 169–189. Birkhäuser Verlag, Basel, Switzerland (preprint available at) http://wslc.math.ist.utl.pt/ftp/pub/CaleiroC/05-CCCM-dyadic.pdf
6. Carnielli, W.A.: Polynomial ring calculus for many-valued logics. In: Werner, B. (ed.) Proceedings of the 35th International Symposium on Multiple-Valued Logic, pp. 20–25. IEEE Computer Society, Los Alamitos (2005), (preprint available at CLE e-Prints vol 5(3), 2005),
   http://www.cle.unicamp.br/e-prints/vol_5,n_3,2005.html
7. Carnielli, W.A., Coniglio, M.E., Marcos, J.: Logics of Formal Inconsistency. In: Gabbay, D., Guenthner, F. (eds.) Handbook of Philosophical Logic, 2nd edn. vol. 14, Kluwer Academic Publishers, Dordrecht (2005), (in print. preprint available at CLE e-Prints vol 5(1), 2005),
   http://www.cle.unicamp.br/e-prints/vol_5,n_1,2005.html
8. Chuang, I.L., Nielsen, M.A.: Quantum Computation and Quantum Information. Cambridge University Press, Cambridge (2000)
9. Clote, P., Kranakis, E.: Boolean Functions and Computation Models. Springer, Heidelberg (2002)
10. Cook, S.A.: The complexity of theorem proving procedures. In: Proceedings of the Third Annual ACM Symposium on the Theory of Computing, pp. 151–158. ACM Press, New York (1971)
11. Copeland, J.: Hypercomputation. Minds and machines 12, 461–502 (2002)
12. Davis, M.: The myth of hypercomputation. In: Teuscher, C. (ed.) Alan Turing: Life and Legacy of a Great Thinker, pp. 195–212. Springer, Heidelberg (2004)
13. Davis, M.: Why there is no such discipline as hypercomputation. Applied Mathematics and Computation 178, 4–7 (2004)
14. Genovese, M.: Research on hidden variable theories: A review of recent progresses. Physics Reports 413, 319–396 (2005)
15. Halpern, J.Y., Harper, R., Immerman, N., Kolaitis, P.G., Vardi, M.Y., Vianu, V.: On the unusual effectiveness of logic in computer science. The Bulletin of Symbolic Logic 7(2), 213–236 (2001)
16. Jacobson, N.: Basic Algebra I, 2nd edn. W. H. Freeman and Company, New York (1985)
17. Papadimitriou, C.: Computational Complexity. Adisson-Wesley, London, UK (1994)
18. Turing, A.M.: On computable numbers, with an application to the Entscheidungsproblem. In: Proceedings of the London Mathematical Society, pp. 230–265 (1936) (A correction, ibid, vol 43, pp. 544–546, 1936-1937)

Amoeba-Based Nonequilibrium Neurocomputer Utilizing Fluctuations and Instability

Masashi Aono and Masahiko Hara

Local Spatio-temporal Functions Lab., Frontier Research System,
RIKEN (The Institute of Physical and Chemical Research),
Wako, Saitama 351-0198, Japan
`masashi.aono@riken.jp`

Abstract. We employ a photosensitive amoeboid cell known as a model organism for studying cellular information processing, and construct an experimental system for exploring the amoeba's processing ability of information on environmental light stimuli. The system enables to examine the amoeba's solvability of various problems imposed by an optical feedback, as the feedback is implemented with a neural network algorithm. We discovered that the amoeba solves the problems by positively exploiting fluctuations and instability of its components. Thus, our system works as a neurocomputer having flexible properties. The elucidation of the amoeba's dynamics may lead to the development of unconventional computing devices based on nonequilibrium media to utilize fluctuations and instability.

Keywords: Bottom-up technology, Physarum, Optimization, Chaos.

1 Introduction

1.1 Fluctuations and Instability

Conventional computing devices are composed of rigid and stable materials to pursue the accuracy and speed of their operations. Fluctuations and instability of the materials are negative disturbing factors that should be suppressed. In contrast, biological systems perform information processing based on fluctuated and unstable materials. Nevertheless, biological organisms succeed in surviving in harsh environments. They often demonstrate their ability to make flexible decisions and seek creative solutions to overcome critical situations. It is difficult for conventional digital computers to demonstrate the flexibility and creativity. Exploring possible forms of processing based on fluctuated and unstable materials, therefore, may lead to the development of flexible and creative computers. This expectation is one of our motives to carry out the study in this paper.

1.2 Nanotechnology for Next-Generation Device

Another motive is to tackle an issue faced by a specific field in nanotechnology researches, where the development of next-generation computing devices integrating a massive amount of molecular-scale components is one of immediate

S.G. Akl et al.(Eds.): UC 2007, LNCS 4618, pp. 41–54, 2007.

priorities. Because it is extremely difficult to precisely manipulate all the components in a top-down manner, bottom-up approaches to utilize self-assembling behavior of the interacting components become increasingly important [1]. At present, in most cases self-assembly is applied in fabrication processes of spatially ordered structures of the components finally immobilized at static equilibrium. The next issue, remaining as an open question, is what to process with the fabricated structure, and how to make its components operate for the processing. The top-down manipulation of the components would be difficult here again. Therefore, it is necessary to explore potentials of processing realizable in a bottom-up manner.

1.3 Self-organization in Nonequilibrium Media

On the other hand, in a field studying nonlinear dynamic systems, there are several proposals to create processing units using self-organized spatiotemporal patterns of chemical components at dynamic nonequilibrium [2]. Some oscillatory or excitable chemical media, such as the Belousov-Zhabotinsky reaction system, generate reaction-diffusion waves. The propagating wave transfers information (for example an ON-state signal) from one site to another. Based on the control of the passage, collision, coincident initiation and annihilation of multiple propagating waves at particular sites, some authors have experimentally demonstrated various processing units, such as a maze solver [3] and logical operators [4]. There is a view that this form of processing gives an extremely simplified model of information processing performed in biological cells.

In this form of processing, the fluctuations and instability in chemical media are vital for providing the background for processing. The randomness driving collisions of reactants is responsible for chemical reactions, and the unstable reaction dynamics is necessary for producing the spatiotemporal patterns. However, unless the fluctuations and instability are kept at a moderate level, they become negative disturbing factors, just as they are suppressed in conventional devices. Indeed, because the coincidence of multiple propagating events is crucial for ensuring the accuracy of processing results, the timing and orbit of the wave propagation should not be fluctuated and destabilized. Some form of top-down precise control of the propagating events is needed for ensuring the accuracy, but it might undermine the significance of utilizing the bottom-up behavior.

The basic principle of the above proposals, that operations of processing units should be accurate and stable, is essentially identical with that of conventional devices. It is necessary to dilute this basic principle for exploring potentials of processing that utilize the bottom-up behavior involving fluctuations and instability. We take a heuristic approach to discover a novel form of processing whose principle is unknown. That is, first we construct a device using a specific medium, and then try to find out what kind of processing would be the most appropriate for the medium. In this paper, we employ a unicellular amoeboid organism that is a kind of biological oscillatory media. As explained in the next section, the amoeba is one of the most suitable organisms for our study.

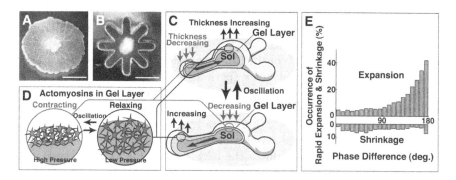

Fig. 1. (A) A huge individual of unicellular amoeba (scale bar=5mm). (B) Au-coated plastic barrier structure on agar plate without nutrients (scale bar=2 mm). The amoeba acts only inside the structure where agar is exposed, as it averts the metal surface. Initial configuration $< 0, 0, 0, 0, 0, 0, 0, 0 >$ is input by placing the amoeba at the center. (C) Schematic of the body structure. (D) Schematic of the contraction-relaxation of actomyosins. (E) The occurrence of rapid expansion/shrinkage of a branch vs. the oscillation phase difference between the branch and hub part. Results for rapid shrinkage are shown upside down.

2   Materials and Methods

2.1   True Slime Mold

An amoeba of true slime mold *Physarum polycephalum* is a huge multinucleated unicellular organism (Fig. 1A). In order to evaluate the amoeba's processing ability by its rationality of the shape deformation, we confine an individual amoeba within the stellate barrier structure put on an agar plate (Fig. 1B). The amoeba concurrently expands or shrinks its multiple branches inside the structure. We employed the amoeba because it has favorable features in its structure and function that fit for our purpose.

Body Structure. The amoeba's body structure is extremely simple and homogeneous. Fig. 1C schematically shows the structure of when there are three expanding branches. An individual amoeba has only a single gel layer (a sort of cellular membrane) enveloping intracellular sol. The gel layer is formed by a massive amount of actomyosins (fibrous proteins contained in muscle) capable of taking contracting or relaxing states (Fig. 1D). These actomyosins are the driver elements of the amoeba, as they play a major role in the shape deformation. Because numerous cell nuclei are distributed throughout the homogeneous body of an individual amoeba, a part of the amoeba divided from the individual survives as another self-sustainable individual. Conversely, multiple individuals fuse together to form an individual when they are contacted. The simplicity and homogeneity of the structure would make the elucidation of the amoeba's dynamics comparatively easy.

Oscillation. The gel layer has a sponge-like property in which extruding and absorbing of intracellular sol is driven by the contraction and relaxation of crosslinked actomyosins at a local site, respectively. Collectively entrained actomyosins generate rhythmic contraction-relaxation cycle observed as vertical thickness oscillation of the gel layer (period=1 $\sim$ 2 min) with various spatiotemporal patterns, where its physiological mechanism has not been elucidated completely. As the phase of the oscillation varies from site to site, intracellular sol is forced to stream horizontally (velocity=$\sim$ 1 mm/sec) by the pressure difference (gradient) derived from the contraction tension difference of the gel layer. The oscillation induces the iterations of direction reversals of the sol streaming in a shuttlewise manner. Intracellular substances including actomyosins and various chemicals essential for the contraction-relaxation oscillation (e.g. Ca^{2+} and ATP) are transported by the sol to travel throughout the whole body.

Deformation. A branch expands or shrinks at a velocity of at most 1 cm/h, as the shuttlewise sol efflux-influx for the branch is iterated for several periods of the oscillation. During our experiment, the amoeba deforms its shape without changing its total volume by keeping a nearly constant amount of intracellular sol. Fig. 1E shows the occurrence rate of rapid expansion/shrinkage of a branch as a function of the oscillation phase difference between the branch and the hub part. When a branch expands rapidly, it is notable that the branch and the hub part synchronize in antiphase (i.e., phase difference $\simeq$ 180°). Conversely, in-phase synchronization (i.e., phase difference $\simeq$ 0°) cannot yield rapid expansion. The velocity of shrinkage was lower than that of expansion, and no significant correlation with the phase difference was confirmed. These results suggest that; i) a branch expands rapidly when its relaxing gel layer invites a greater influx of the sol extruded by the contracting gel layer of the hub part (Fig. 1C), and ii) the shrinkage of a branch is driven passively as the sol flows out to other expanding branches. Namely, the amoeba's decision on its horizontal shape deformation is made by spatiotemporal patterns of vertical thickness oscillation [5].

Optimization Capability. Despite the absence of a central system, the amoeba exhibits sophisticated computational capacities in its shape deformation. Indeed, the amoeba is capable of searching for a solution of a maze [6]. Inside a barrier structure patterned to have walls of the maze, the amoeba deforms its shape into a string-like configuration that is the shortest connection path between two food sources at the entrance and exit of the maze. This result shows that the amoeba realizes the optimization of the nutrient absorption efficiency and leads its whole body to a favorable condition for the survival. The amoeba's shape deformation is handled by spatiotemporal oscillation patterns. The spatiotemporal pattern is self-organized by collectively interacting actomyosins. Therefore, the amoeba can be regarded as a sophisticated information processor utilizing the bottom-up behavior of massively parallel components. This feature fits for exploring potentials of processing realizable with nonequilibrium media in a bottom-up manner.

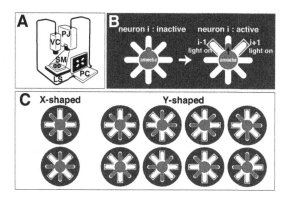

Fig. 2. (A) Optical feedback system. For transmitted light imaging using a video camera (VC), the sample circuit (SM) was illuminated from beneath with a surface light source (LS). The recorded image was processed using a PC to update the monochrome image for illumination with a projector (PJ). (B) Optical feedback rule: active state $x_i(t) = 1$ triggers light illumination $y_{i-1}(t + \Delta t) = y_{i+1}(t + \Delta t) = 1$ (white light projected to rectangular regions). (C) All possible solutions of the problem.

2.2 Neural Network Implementation

State of Neuron. We call the i th path of the stellate structure "neuron i" ($i \in I = \{1, 2, \cdots, N = 8\}$). The amoeba's shape is monitored using a video camera at each interval of $\Delta t = 6$ sec (Fig. 2A). For each neuron i at time t, whenever more than a quarter of the area of the i th neuron is occupied by the amoeba's branch, state 1 (active) is assigned as $x_i(t) = 1$, otherwise $x_i(t) = 0$ (inactive). The amoeba's shape is freely deformable to fit into arbitrary boundary condition of any network topology. Thus, the computing can be started from arbitrary input configuration $< x_1(0), x_2(0), \cdots, x_N(0) >$.

Photoavoidance-Based Control. The amoeba's branch exhibits photoavoidance response. A branch is shrunk by a local light stimulation, as the light-induced contraction enhancement of the gel layer intensifies the sol efflux (extrusion) from the stimulated region [7]. A neuron, therefore, is led to be inactive by illuminating the corresponding region. The illumination is carried out with a projection of a monochrome image such that only the regions to be illuminated are colored in white (Fig. 2A). We write $y_i = 1$ if the light for the neuron i is turned on. Conversely, if the neuron i is allowed to be $x_i = 1$, the light is turned off as $y_i = 0$. Any neuron is activated naturally if it is not illuminated, because the amoeba inherently tries to expand all branches to generate the configuration $< 1, 1, 1, 1, 1, 1, 1, 1 >$. The illumination pattern $< y_1, y_2, \cdots, y_N >$, therefore, leads the system configuration $< x_1, x_2, \cdots, x_N >$ toward the ideal configuration $< x_1', x_2', \cdots, x_N' >$ such that all neurons satisfy the follwing condition: $x_i' = 1 - y_i$. We call this condition "counteractive rule" and adopt it as our system's control principle.

Feedback Dynamics. In assuming the counteractive rule, the optical feedback system automatically updates the illumination pattern. The feedback is implemented with a discrete form of recurrent neural network dynamics known as Hopfield-Tank model [8]: $y_i(t + \Delta t) = 1 - f(\Sigma_{j=1}^{N} w_{ij} x_j(t))$, where w_{ij} is the weight assigned to the link from the neuron j to neuron i, and the threshold $\theta = 0$ defines the step function $f(\sigma) = 1$ if $\sigma \geq \theta$, otherwise 0. The illumination pattern is updated at each interval with the change in the system configuration, and again it induces further deformation of the amoeba.

Stable Equilibrium Solution. Any configuration is called "stable equilibrium" if and only if the counteractive rule is satisfied for all neurons. In a stable equilibrium, the amoeba is no longer forced to reshape by illumination and can fully expand its branches inside all nonilluminated neurons. Thus, the configuration would be comfortable for the amoeba, and is expected to be stably maintained unless the amoeba spontaneously breaks the counteractive rule. In the formalization of the recurrent neural network, a stable equilibrium is regarded as a solution of the problem defined by assigning the weights. In general, there exist multiple solutions, and each solution has minimum energy of the potential energy landscape established by the network dynamics. Any medium that can only relax toward its stable equilibrium, therefore, cannot bootstrap itself out of a once-reached solution having locally minimum energy.

2.3   Embodiment of Problem Solving

Constraint Satisfaction Problem. We assigned the weights as $w_{ij} = -1$ if $|i - j| = 1$, otherwise 0, to define the following symmetric rule for updating the illumination (Fig. 2B): If the neuron i is active ($x_i(t) = 1$), its adjacent neurons $i - 1$ and $i + 1$ are induced to be inactive by illumination ($y_{i-1}(t + \Delta t) = y_{i+1}(t + \Delta t) = 1$), where the boundary condition is periodic. This rule prohibits two adjacent neurons i and $i+1$ from taking active state simultaneously. Another interpretation of this rule is that each neuron is led to execute logical *NOR*-operation known as a universal logic element: The neuron i is illuminated to be inactive ($x_i(t+\Delta t) = 0$), if at least one of its adjacent neurons is active ($x_{i-1}(t) = 1$ or $x_{i+1}(t) = 1$), otherwise ($x_{i-1}(t) = x_{i+1}(t) = 0$) nonilluminated to be active ($x_i(t + \Delta t) = 1$). Namely, the optical feedback imposes the following constraint satisfaction problem: Find the system configuration $< x_1, x_2, \cdots, x_8 >$ such that all neurons satisfy $x_i = NOR(x_{i-1}, x_{i+1})$. As a solution of this problem, the amoeba is required to search for a stable equilibrium. There are 10 solutions (Fig. 2C) consisting of rotation symmetric groups of $< 1, 0, 1, 0, 1, 0, 1, 0 >$ ("X-shaped") and $< 1, 0, 0, 1, 0, 0, 1, 0 >$ ("Y-shaped").

Livelock and Deadlock. The concurrent processing of circularly connected *NOR*-operators has a potential of creating a stalemated unsolvability of the problem, when all operations are executed in a synchronous manner. Let us suppose that all branches expand or shrink with a uniform velocity. From the initial configuration $< 0, 0, 0, 0, 0, 0, 0, 0 >$ evoking no illumination, the synchronous expanding movements of all branches will lead to $< 1, 1, 1, 1, 1, 1, 1, 1 >$. In this configuration, all neurons are illuminated. Then, all branches shall shrink uniformly to evacuate from the illuminations. This leads all branches to return to the initial configuration that allows them to expand again. In this manner, the system can never attain a solution, as the synchronous movements result in a perpetual oscillation between $< 0, 0, 0, 0, 0, 0, 0, 0 >$ and $< 1, 1, 1, 1, 1, 1, 1, 1 >$. This type of stalemated situation is called "livelock".

On the other hand, let us assume that the problem could be solved successfully. The amoeba is expected to stably maintain the solution because it would be a temporarily comfortable condition. However, the amoeba would inherently intend to search for foods by the deformation with the expanding behavior of branches. Permanently maintaining the solution, therefore, eliminates any possibility of feeding and brings about its death from starvation[1]. This is another type of stalemated situation resembling "deadlock". Thus, our experiment requests the achievement of the following two goals to the amoeba; the amoeba should be capable of not only livelock avoidance for attaining a solution, but also deadlock avoidance for breaking away from the solution [9,10,11]. We carried out the experiment to observe how the amoeba copes with the stalemated situations in a bottom-up manner.

3   Results

3.1   Problem Solving Process

Spatiotemporal Oscillation Pattern. The experiment was started from the initial configuration $< 0, 0, 0, 0, 0, 0, 0, 0 >$ (Fig. 1B) by placing the amoeba's spherical piece (0.75±0.05 mg) at the center of the stellate structure. In the early stage, the amoeba flattened its spherical shape to turn into disc-like shape, and horizontally expanded its thin periphery in a circularly symmetric manner. As shown in panels of Fig. 3, by binarizing the phase of vertical thickness oscillation of each local site into increasing (relaxing) and decreasing (contracting) phases, we can clearly observe spatiotemporal oscillation patterns.

Spontaneous Symmetry-Breaking. The amoeba's circular deformation (i.e., synchronous expanding movements of all branches) would be inevitable if realizable spatiotemporal patterns were limited to circularly symmetric ones. However, we observed that the amoeba produces various symmetry-broken patterns, and the pattern varies nonperiodically in a complex manner. As shown in Fig. 3A, the symmetry was spontaneously broken as a defective site was created in the periphery. The defective site was exclusively expanded at a velocity relatively larger than that of the rest, and developed into a distinct branch having enough occupying area inside the neuron 8 to evoke the illumination.

---

[1] Although no food source is given in our experimental condition, the amoeba in our experiment can survive without any food for up to about a week because it can store nutrients fed before the experiment as internal energy source.

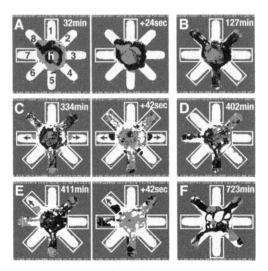

Fig. 3. Time course of events in computing process. (A) Transient configuration $< 0, 0, 0, 0, 0, 0, 0, 1 >$. Illuminated regions are indicated by rectangles. The oscillation phase is binarized into the relaxing (thickness increasing) and contracting (decreasing) states, represented by the black (red in colored version) and gray (blue) pixels, respectively. (B) First-reached solution $< 0, 1, 0, 0, 1, 0, 0, 1 >$ (duration$\simeq$4 h). (C) Spontaneous destabilization. The solution B $< 0, 1, 0, 0, 1, 0, 0, 1 >$ (left) evolved into $< 0, 1, 1, 0, 1, 0, 1, 1 >$ (right). Arrows indicate the growth directions of the newly emerged branches expanding under illumination contrary to photoavoidance. (D) Second-reached solution $< 0, 1, 0, 0, 1, 0, 1, 0 >$ (duration$\simeq$1 h). (E) Spontaneous destabilization. The solution D $< 0, 1, 0, 0, 1, 0, 1, 0 >$ (left) evolved into $< 0, 1, 0, 0, 1, 0, 1, 1 >$ (right). (F) Third-reached solution $< 0, 1, 0, 1, 0, 1, 0, 1 >$ (duration$\simeq$7 h).

Solution Seeking with Livelock-Breaking. Sometimes we observe the livelock-like situation in which the amoeba's circularly symmetric deformation continues by keeping its disc-like shape until all neurons are illuminated. Even in that case, the symmetry is spontaneously broken and movements of peripheral parts become asynchronously fluctuated with mutual time lags. Owing to the asynchronously fluctuated movements, the livelock-like stalemate was avoided, and the amoeba searched for the solution $< 0, 1, 0, 0, 1, 0, 0, 1 >$ shown in Fig. 3B. The solution was stably maintained for about 4 h. We observed that the spatiotemporal pattern of vertical oscillation evolves nonperiodically while the amoeba's shape in the solution is unchanged horizontally [12,13,14].

3.2   Searching Ability of Multiple Solutions

Destabilization of Solution with Deadlock-Breaking. The stabilization of the solution (Fig. 3B) was longly persisted as if the amoeba is stalemated in a deadlock-like situation. However, the stabilizing mode was spontaneously switched to the destabilizing mode without any explicit external perturbation.

Interestingly, the amoeba broke through the deadlock, as two branches suddenly emerged and started to invade illuminated regions (Fig. 3C). The invading behavior against the counteractive rule is contrary to the photoavoidance response.

Spontaneous Switching between Stabilizing and Destabilizing Modes.
While aggressive expansion of the branch 7 was sustained under illumination, the branch 8 was shrunk by illumination. In this manner, the branches iterated their asynchronously fluctuated movements again. As a result, the amoeba succeeded in searching for another solution $< 0, 1, 0, 0, 1, 0, 1, 0 >$ (Fig. 3D). Then amoeba stabilized its shape at the second solution for about 1 h. After that, the spontaneous destabilization occurred once more (Fig. 3E), and eventually the amoeba attained the third solution $< 0, 1, 0, 1, 0, 1, 0, 1 >$ maintained for about 7 h (Fig. 3F). Within 16 h of the observation, the amoeba successively searched for three solutions by repeatedly switching between the stabilizing and destabilizing modes.

3.3   Statistical Results

We carried out 20 experimental trials measured for an average of 14.5 h. As shown in Fig. 4A, in every trial at least one solution was searched, and transition among an average of 3.25 solutions (gross number including multiple attainments of an identical solution) was realized per a trial. This result implies that the amoeba significantly has the capabilities of searching for solutions and destabilizing the solutions.

The transition among solutions was observed as nonperiodic. As shown in Fig. 4B, the duration of a solution distributes within a range of about 0.5 h to 12 h. The durations of X-shaped solutions are relatively longer than that of Y-shaped ones. This suggests that X-shaped solutions are more stable than Y-shaped ones. The transition is stochastic in a sense that a solution has a potential to evolve into multiple solutions. We confirmed that transition from more stable X-shaped solution to less stable Y-shaped one is possible.

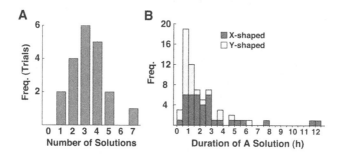

Fig. 4. Statistical results. All data were obtained from 20 trials. (A) Frequency distribution of the number of solutions searched. (B) Frequency distribution of the duration time of stay in a solution. Results were evaluated separately for X-shaped and Y-shaped solutions, and are shown in a stacked manner (e.g., Y-shaped ones maintained for the range of 0.5 to 1 h were observed 13 times).

4   Discussions

4.1   Biological Implication of Spontaneous Destabilization

In the stabilizing mode, all branches keep away from illumination for maintaining a solution. In the destabilizing mode, under equal illumination conditions, some branches enterprisingly invade aversive illuminated regions contrary to their photoavoidance, while others remain shrunk as usual. This implies that the amoeba's photoavoidance response is nonuniform spatially and temporally. This flexibly variable stimulus-response would be essential for the survival of the amoeba required to search for food in a harsh environment, because it enables the amoeba surrounded by aversive stimuli to break through the stalemated situation with enterprising responses. In our experiment, the deadlock-like critical situation in which the amoeba is stuck in its starved condition was flexibly broken by the amoeba's spontaneous destabilization. That is, unstable phenomena have a positive aspect to produce the flexibility.

4.2   Mechanism of Emergence of New Branch

The spontaneous destabilization is realized as new branches emerge suddenly, although its mechanism remains to be seen. In most cases, a new branch emerges as a singularly thick structure with localized large vertical oscillation amplitude. It appears to intensively invite a drastic sol influx from a broad domain. Figure 5 illustrates our speculation that the singularly thick structure is formed as intracellular sol spouts vertically from the gel layer's defective site where it is locally fragile to be ruptured (i.e., crosslinking of actomyosins is broken). According to this assumption, the structure of the newly emerged branch is composed chiefly of extruded sol and is not covered by a sufficient quantity of gel to actualize light-induced contraction. Such a soft blob-like structure must be incapable of photoavoidance response, and thus it can be extruded to expand flexibly even under unfavorable illuminated conditions. Fragile and unstable components of biological systems may provide a positive aspect to enable flexible behavior.

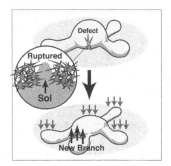

Fig. 5. Schematic of speculated process in the emergence of a new branch. A fragile defect of the gel layer is ruptured, and a new branch is formed as the sol spouts from the defect. The branch rapidly develops through self-repairing process of the defect.

4.3   Origin of Spontaneous Destabilization

The spontaneous switching between the stabilizing and destabilizing modes (i.e., the transition among solutions) was observed as nonperiodic and stochastic. As one of its possible origins, we suppose the combination of intrinsic microscopic fluctuations and spatiotemporal chaos characterized as nonperiodic but nonrandom behavior. Indeed, the system's behavior is chaotic as its time evolution is unstable and unreproducible. Its capability of successively searching for multiple solutions, however, is robustly maintained and qualitatively reproducible. This resembles the robustness of strange attractors of chaotic systems.

Unstable chaotic dynamics with properly tuned parameter intermittently switches between the stabilizing and destabilizing modes by exponentially amplifying microscopic fluctuations [15]. That is, fluctuations in the microscopic level such as the thermal noise, nonuniformity of chemical distributions, and the gel layer's uneven stiffness distribution (density distribution of crosslinked actomyosins), are amplified to an extensive and sustained movements of the macroscopic level. Some form of positive feedback effect by the coupling of chemical, physiological, and hydrodynamic processes may be responsible for the amplification process.

4.4   Applications

Logical Circuit, Associative Memory, and Optimization Problem Solver. The amoeba's capability to search for a stable equilibrium under our optical feedback allows our system to be developed to a logical circuit holding a kind of computational universality. By properly altering the system parameters including the number of neurons, thresholds, and weights, our system can implement a network of the McCulloch-Pitts neurons capable of simulating any Boolean logic operations with the network architecture allowing arbitrary circuit wiring [16]. In addition, the system can be developed to an associative memory device by regarding a stable equilibrium as a predefined memory pattern.

Moreover, the amoeba's capability of transition among multiple stable equilibria may fit for solving combinatorial optimization problem (e.g. traveling salesman problem), because it may assist searching for a global optimum solution without being stuck at local optima. The usefulness of chaotic dynamics for combinatorial optimization has already been clarified with chaotic neural network models, where both stabilizing and destabilizing effects contribute to efficient searching dynamics [17,18].

Future Perspective on Flexible Device. In the conventional computing paradigm, deadlock can be avoided in the software level, if it is possible to prescribe an appropriate resource allocation protocol comprehending potential resource requests from all processes. However, no programmer can know what the processes will request in advance of the coding without actual executions. This means that deadlock avoidance is often impossible for many practical systems. On the other hand, our system flexibly breaks through the livelock-like and deadlock-like situations and searches for several solutions without providing any particular resource allocation protocol. This is because the amoeba's spatiotemporal oscillation patterns autonomously materialize and alter the allocation of intracellular sol to its branches in a bottom-up manner. In contrast to the conventional paradigm based on hardwares operating passively as instructed by software programs, the amoeba as a hardware in our system actively alters the interpretation of the software program (i.e., the control principle with the feedback algorithm). This unique capability would be advantageous in the development of autonomous systems operated in actual environments, such as robot control systems [19]. It may enable the robot to flexibly respond to concurrent occurrences of unexpected events that would lead the robot to stalemated situations, even if the programs prescribed only for expected events were defective or useless for searching for solutions to overcome the situations.

Although dramatic improvement in its slow processing speed is unpromising as long as the amoeba is used, it may be possible to develop our computing scheme by implementing with other faster materials, if the dynamics of the amoeba's oscillatory behavior is elucidated. Potential candidates suitable as the alternative materials would be oscillatory or excitable nonequilibrium media capable of spontaneously breaking the spatiotemporal symmetries in a chaotic manner.

5 Conclusions

Our discoveries and discussions in this paper are summarized as follows.

1. Fluctuations and instability in the amoeba's components have positive aspects to produce the enhancement of searching ability and the flexibility of decision-making ability.
   - The amoeba is capable of searching for a stable equilibrium solution of a constraint satisfaction problem, due to its ability to fluctuate its movements for breaking through a livelock-like stalemated situation.
   - The amoeba is capable of searching for multiple solutions one after another by spontaneously destabilizing once-reached solutions, because it has flexibility to change its stimulus-response for breaking through a deadlock-like stalemated situation.
2. The above problem solving is realizable owing to the amoeba's rebellious behavior against the feedback control, and it presents a possible form of processing to utilize the bottom-up behavior of fluctuated and unstable materials.
3. The amoeba's searching ability and flexibility allow our system to be developed to logical circuits, associative memory devices, combinatorial optimization problem solvers, and novel devices that flexibly respond to occurrences of unexpected events.
4. If the amoeba's oscillatory dynamics is elucidated, our proposal may be implementable by other faster nonequilibrium medium.

References

1. Whitesides, G.M., Grzybowski, B.A.: Self-Assembly at all Scales. Science 295, 2418–2421 (2002)
2. Adamatzky, A., De Lacy Costello, B., Asai, T.: Reaction-Diffusion Computers. Elsevier, UK (2005)
3. Steinbock, O., Toth, A., Showalter, K.: Navigating Complex Labyrinths: Optimal Paths from Chemical Waves. Science 267, 868–871 (1995)
4. Motoike, I., Yoshikawa, K.: Information operations with an excitable field. Phys. Rev. E 59, 5354–5360 (1999)
5. Nakagaki, T., Yamada, H., Ueda, T.: Interaction between cell shape and contraction pattern. Biophys. Chem. 84, 195–204 (2000)
6. Nakagaki, T., Yamada, H., Toth, A.: Maze-solving by an amoeboid organism. Nature 407, 470 (2000)
7. Ueda, T., Mori, Y., Nakagaki, T., Kobatake, Y.: Action spectra for superoxide generation and UV and visible light photoavoidance in plasmodia of Physarum polycephalum. Photochem. Photobiol. 48, 705–709 (1988)
8. Hopfield, J.J., Tank, D.W.: Computing with neural circuits: A model. Science 233, 625–633 (1986)
9. Aono, M., Gunji, Y.P.: Beyond input-output computings: Error-driven emergence with parallel non-distributed slime mold computer. BioSystems 71, 257–287 (2003)
10. Aono, M., Gunji, Y.P.: Resolution of infinite-loop in hyperincursive and nonlocal cellular automata: Introduction to slime mold computing. In: Dubois, D.M. (ed.) Computing Anticipatory Systems: CASYS 2003. AIP conference proceedings, vol. 718, pp. 177–187 (2004)
11. Aono, M., Gunji, Y.P.: Material implementation of hyperincursive field on slime mold computer. In: Dubois, D.M. (ed.) Computing Anticipatory Systems: CASYS 2003. AIP conference proceedings, vol. 718, pp. 188–203 (2004)
12. Takamatsu, A., et al.: Time delay effect in a living coupled oscillator system with the plasmodium of Physarum polycephalum. Phys. Rev. Lett. 85, 2026 (2000)
13. Takamatsu, A., et al.: Spatiotemporal symmetry in rings of coupled biological oscillators of Physarum plasmodial slime mold. Phys. Rev. Lett. 87, 078102 (2001)
14. Takamatsu, A.: Spontaneous switching among multiple spatio-temporal patterns in three-oscillator systems constructed with oscillatory cells of true slime mold. Physica D 223, 180–188 (2006)
15. Kaneko, K., Tsuda, I.: Complex Systems: Chaos and Beyond - A Constructive Approach with Applications in Life Sciences. Springer, New York (2001)
16. Arbib, M.A. (ed.): The Handbook of Brain Theory and Neural Networks. The MIT Press, Cambridge, Massachusetts (2003)
17. Aihara, K., Takabe, T., Toyoda, M.: Chaotic Neural Networks. Phys. Lett. A 144, 333–340 (1990)
18. Hasegawa, M., Ikeguchi, T., Aihara, K.: Combination of Chaotic Neurodynamics with the 2-opt Algorithm to Solve Traveling Salesman Problems. Phys. Rev. Lett. 79, 2344–2347 (1997)
19. Tsuda, S., Zauner, K.P., Gunji, Y.P.: Robot Control with Biological Cells. In: Proceedings of Sixth International Workshop on Information Processing in Cells and Tissues, pp. 202–216 (2005)

Appendix

Experimental Setups

The amoeba was fed by oat flakes (Quaker Oats, Snow Brand Co.) on a 1% agar gel at 25°C in the dark. The stellate barrier structure (thickness approximately 0.2 mm) is made from an ultrathick photoresist resin (NANO^{TM} XP SU-8 3050, MicroChem Corp.) by a photolithography technique, and was coated with Au using a magnetron sputter (MSP-10, Shinkuu Device Co., Ltd.). The experiments were conducted in a dark thermostat and humidistat chamber (27±0.3 °C, relative humidity 96±1%, THG062PA, Advantec Toyo Kaisha, Ltd.). For transmitted light imaging, the sample was placed on a surface light guide (MM80-1500, Sigma Koki Co., Ltd.) connected to a halogen lamp light source (PHL-150, Sigma Koki Co., Ltd.) equipped with a band-pass filter (46159-F, Edmund Optics Inc.), which was illuminated with light (intensity 2 μW/mm^2) at a wavelength of 600±10 nm, under which no influence on the amoeba's behavior was reported (Ueda et al., 1988). The intensity of the white light (monochrome color R255:G255:B255) illuminated from the projector (3000 lm, contrast ratio 2000:1, U5-232, PLUS Vision Corp.) was 123 μW/mm^2. The outer edge of the circuit (border between the structure and the agar region) was always illuminated to prevent the amoeba from moving beyond the edge.

Unconventional "Stateless" Turing–Like Machines

Joshua J. Arulanandham

School of Computing Sciences
VIT University, Vellore, India
hi_josh@hotmail.com

Abstract. We refer to Turing machines (TMs) with only one state as "stateless" TMs. These machines remain in the same state throughout the computation. The only way they can "remember" anything is by writing on their tapes. Stateless TMs have limited computing power: They cannot compute all computable functions. However, this handicap of a stateless TM can be overcome if we modify the TM a little by adding an extra feature. We propose a modified Turing–like machine called a JTM—it is stateless (has only one internal state) by definition— and show that *it is as powerful as any TM*. That is, a JTM does not switch states, and yet can compute all computable functions. JTMs differ from conventional TMs in the following way: The tape head spans three consecutive cells on the tape; it can read/write a *string* of three (tape) symbols all at once. However, the movement of the head is not in terms of "blocks of three cells": in the next move, the head does not "jump" to the adjacent (entirely new) block of three cells; it only *shifts* itself by one cell—to the right or to the left. A JTM is more than a product of theoretical curiosity: it can serve as a "normal form" and might lead to simpler proofs for certain theorems in computation theory.

1 Turing Machines

Turing machines (TMs), proposed by Alan Turing [2] in 1936, are one of the first "toy models" that were used to define precisely the notions of *computation* and *computability*—what it means to "compute", and what can and cannot be computed. A TM is actually a simple mechanical device that does symbol–manipulation (which one can actually build, if one wants to!), but is often employed as a "paper machine", i.e. as a theoretical construct in computer science. Theorems regarding the nature of computation proven using TMs apply to real computers, since TMs faithfully model real computers, i.e. any input–output function computable on a digital computer can be computed on a TM [1]. In what follows we first give an informal description of the Turing machine, and then, some formal definitions.

The Turing machine has a tape that is divided into cells (squares); the tape extends infinitely to the right and to the left. Each cell of the tape can hold a symbol. The tape initially contains the *input*—a finite string of symbols in the

S.G. Akl et al.(Eds.): UC 2007, LNCS 4618, pp. 55–61, 2007.

input alphabet—and is also used to hold the *output*, once the computation is over. Cells other than those containing the input are initially "blank", i.e. they hold the special (blank) symbol B. "Computation" on a TM is accomplished by reading and (over)writing symbols on the tape: There is a tape head (initially positioned over the left most symbol of the input) that hops from cell to cell, reading and writing a symbol. The tape head (its movement, read/write process) is controlled by a control unit that can switch between different *states* (a finite set), each of which represents a certain "mode of operation"[1]. Each *move* of a TM depends on two things: the state of the control unit and the symbol being read (scanned) by the tape head. In one move, a TM will: (i) *Write* a (possibly, new) symbol on the cell under the tape head, (ii) *Shift* the tape head one cell, to the left or to the right and, (iii) *Update the state* of the control unit, possibly changing it to a new state.

Definition 1. *A TM can be described as a 6–tuple $M = (Q, \Sigma, \Gamma, \delta, q_0, B)$ where Q is the finite set of* states *of the control unit, Σ is the finite set of input symbols, Γ is the set of tape symbols (Σ is a proper subset of Γ), $\delta : Q \times \Gamma \to Q \times \Gamma \times \{L, R\}$ is the transition function, $q_0 \in Q$ is the* start *state of the control unit, and B is the blank symbol (which is in Γ, but not in Σ) that initially appears throughout the tape except on finitely many cells that contain the input symbols. The transition function $\delta(q, X)$ takes as arguments a state q and a tape symbol X and returns a triple (q', X', D) where $q' \in Q$ is the next state, $X' \in \Gamma$ is the symbol written on the cell being scanned, and D is a direction (left or right) that tells where the head shifts in the next move.*

The state of the control unit, the symbol being scanned on the tape, and the entire content of the tape together constitute the *configuration* ("snapshot") of a TM and will be described with the string $X_1 X_2 \ldots X_{i-1} q X_i X_{i+1} \ldots X_n$, where q is the state of the machine, $X_1 X_2 \ldots X_n$ is the portion of the tape between the leftmost and rightmost nonblanks and, X_i is the symbol being scanned.

A TM will *halt* if it reaches some state q while scanning a certain tape symbol X, and no next move is defined for this situation, i.e. $\delta(q, X)$ is left undefined. When a machine halts, the portion of the tape between the leftmost and rightmost nonblanks represents the *output* of the computation.

Finally, we offer an idea that is useful while programming a TM: It is sometimes easier if we imagine the tape as being composed of multiple tracks, each track containing its own sequence of symbols. Thus, in the "new" multi–track machine, sequence of symbols appear one over the other, and therefore, a single cell must be thought of as holding a "mega symbol"—a tuple such as $[X, Y, Z]$, for example (the tape alphabet will consist of tuples). Note that the multi–track idea has *not* changed the basic Turing machine model in any way.

The focus of this paper is on TMs with only one state, which we refer to as "stateless" TMs. These machines remain in the same state throughout the computation. The only way they can "remember" anything is by writing on their tapes. Stateless

---

[1] Restricting the number of states of a TM to one restrains its computational power. TMs with only one state cannot compute all computable functions.

TMs have limited computing power: They cannot compute all computable functions. However, this handicap of a stateless TM can be overcome if we modify the TM a little by adding an extra feature. In the next section, we introduce a modified Turing–like machine called a JTM—it is stateless (has only one internal state) by definition—and show that *it is as powerful as any TM*. That is, a JTM does not switch states, and yet can compute all computable functions.

2   Enter JTM

A JTM is a Turing machine (as defined in the previous section) with the following special features:

- The tape head is wider than that of an ordinary TM. It spans three consecutive cells on the tape (we will refer to the three cells being scanned as the current "window"); it can read/write a *string* of three (tape) symbols under the window, all at once[2]. However, the movement of the head is not in terms of "blocks of three cells": in the next move, the head does not "jump" to the adjacent (entirely new) block of three cells; it only *shifts* itself by one cell—to the right or to the left.
- The finite state control has only one state and remains in that state forever. This means, the notion of "state switching" (and consequently, the very notion of "states") becomes unnecessary in the context of a JTM.

Note that, initially (before the computation begins), the tape head will be scanning the leftmost cell containing the input and the two cells on its right. A move of the JTM is a function of only one thing: the 3–letter string in the window being scanned. Based on this 3–letter string every "next move" of the JTM will:

(a) First, *write* a 3–letter string (of tape symbols) in the window being scanned.

(b) Then, *shift* the tape head by one cell—to the left or to the right.

We now give a formal definition of a JTM.

Definition 2. *A JTM can be described as a 4–tuple $J = (\Sigma, \Gamma, \delta, B)$ where Σ is the finite set of input symbols, Γ is the set of tape symbols (Σ is a proper subset of Γ), $\delta : \Gamma^3 \to \Gamma^3 \times \{L, R\}$ is the transition function, and B is the blank symbol (which is in Γ, but not in Σ) that initially appears throughout the tape except on finitely many cells that contain the input symbols. The transition function $\delta(X)$ maps X, the 3–letter string over Γ in the window being scanned, to a pair (X', D) where X' is a 3–letter string that overwrites the string in the window, and D is a direction (left or right) that tells where the head shifts in the next move.*

Observe that, in the above definition of a JTM, the components Q (set of states) and q_0 (initial state) have been dropped from the 6–tuple definition of a TM.

---

[2] The usage of the phrase "current window" in the context of a JTM is analogous to that of "current cell" in a TM.

Like multi–track TMs, we can have JTMs with multiple tracks where each tape symbol is a tuple. In the following section we will use a two–track JTM to simulate a TM.

3   JTMs Can Simulate TMs

The question arises: Why is a new TM incarnation necessary at all? The answer is, when we "relax" the definition of a TM by allowing the head to scan a *window* of three cells (as opposed to scanning a single *cell*), a miracle occurs: An explicit notion of the *state* of the finite control becomes an extra baggage that can be gotten rid of. In a sense, it becomes easier and simpler to describe what goes on inside a JTM during each "move": The head reads/writes a string and makes a shift in a given direction. That is all. The state changes in the finite control need no longer be tracked in the new model. And, all this comes without having to lose any of the original computing power of a TM, a fact we will formally prove in what follows.

Theorem 1. *Any input–output function computable by a Turing machine can be computed by a JTM.*

Proof. Consider an input–output function $f : \Sigma^* \to \Sigma^*$ that is computable by a TM $M = (Q, \Sigma, \Gamma, \delta, q_0, B)$, where the symbols have the usual meanings (as discussed in the preceding sections). We will directly construct a JTM J with two tracks that simulates M on input w, thus proving that JTM J can compute f. Let $J = (\{B\} \times \Sigma, (\{B\} \cup Q) \times \Gamma, \delta', [B, B])$ where δ' is the transition function which will be described shortly. From now on we will write a symbol such as $[B, B]$ as $\genfrac{}{}{0pt}{}{B}{B}$, the way it would actually appear on the two–track tape. Initially J holds on the lower track of its tape whatever M would have on its tape initially, i.e. the input w surrounded by blanks (B's). The upper track, which is completely blank to start with, will be used to keep track of the simulated states of M: J will "scribble" state labels (such as "q_0", "q_1", etc.) on the upper track so as to remember which state M is currently in. (*M*'s simulated states are recorded in such a way that *M 's current state and the current symbol being scanned will always appear together as a single symbol (tuple) on the same cell of the tape.*) Before the simulation begins, the tape head is parked over the leftmost cell holding the input and the two cells immediately on its right.

·   In this paragraph, before formally spelling out the transition rules of J, we give an inkling of how the transition rules will be framed. In what follows we will often refer to three distinct sections of the window (which, in a 2–track JTM, resembles a 2×3 matrix) scanned by J's tape head: the "left column", the "middle column" and the "right column" of the window. We will "program" J in the following way: While simulating each move of M on w, *the middle column of the window (currently being scanned) will always hold M's simulated state, say q, (on the upper track) and the symbol currently being scanned by M, say b (on the lower track).* Having q and b under the "view" of the tape head (in the middle column[3] of the current window, as in $\overset{x}{a}\ \overset{\mathbf{q}}{\mathbf{b}}\ \overset{y}{c}$, for example), it is easy for J to compute/simulate M's next move: Simply simulate M's transition rule applicable in this "situation", i.e. $\delta(q, b) = (q', b', L)$ (say), where L denotes left direction of the tape head. How? By overwriting the current window (which is $\overset{x}{a}\ \overset{\mathbf{q}}{\mathbf{b}}\ \overset{y}{c}$ in our example) with an appropriate 3–letter string (say, with $\overset{q'}{a}\ \overset{q}{b'}\ \overset{y}{c}$) that produces the following effect, and after that, shifting the tape head one cell to the left:

(i) b gets replaced with b' as required, and
(ii) q' (the simulated next state of M) is written on the *left* column (upper track) of the window, exactly above the next symbol about to be scanned. (After the left shift, the *middle* column of the "new" window will be exactly over $\overset{q'}{a}$, as required.)

We are now ready to describe the transition function δ' of J. Recall that $\delta'(X)$ maps X, a 3–letter string in the window being scanned (which would look like $\overset{B}{a}\ \overset{B}{b}\ \overset{B}{c}$, since J has two tracks on its tape), to a pair (X', D) where X' is a 3–letter string that overwrites the string in the window, and D is a direction (left or right) that tells where the head shifts in the next move.

1. For all symbols $a, b, c \in \Gamma$ frame the following rule for J:

$$\delta'(\overset{B}{a}\ \overset{\mathbf{B}}{\mathbf{b}}\ \overset{B}{c}) = (\overset{q_0}{a}\ \overset{B}{b}\ \overset{B}{c}, L)$$

The above rule will be invoked only once (in the beginning) during our simulation. This is a preprocessing step (0th move of J) which writes q_0, the initial state of M, on the upper track above the first symbol of input w and shifts the head to the left.

2. For each rule

$$\delta(q, b) = (q', b', L)$$

in M, where $q, q' \in Q$; $b, b' \in \Gamma$ and L denotes left direction, frame the following transition rule for J:

$$\delta'(\overset{x}{a}\ \overset{\mathbf{q}}{\mathbf{b}}\ \overset{y}{c}) = (\overset{q'}{a}\ \overset{q}{b'}\ \overset{y}{c}, L)$$

for all $a, c \in \Gamma$ and, for all $x, y \in \{B\} \cup Q$.

3. For each rule

$$\delta(q, b) = (q', b', R)$$

in M, where $q, q' \in Q$; $b, b' \in \Gamma$ and R denotes right direction, frame the following transition rule for J:

$$\delta'(\overset{x}{a}\ \overset{\mathbf{q}}{\mathbf{b}}\ \overset{y}{c}) = (\overset{x}{a}\ \overset{q}{b'}\ \overset{q'}{c}, R)$$

for all $a, c \in \Gamma$ and, for all $x, y \in \{B\} \cup Q$.

---

[3] The other two columns do not influence M's simulated next move.

What remains now is to prove that the above construction is, indeed, correct. We will prove that the machines M and J will have the same configuration after each of them makes n moves, for any $n \geq 0$. (This would automatically imply that both of them produce the same output for the same input.) Of course, M may not go on forever and would eventually halt if, at some point, no "next move" is defined; J would follow suit, since J too would not have a (corresponding) rule that dictates its next move. (Note that we do not count the initial move of J as the "first move", but as the "zeroth move"—a preprocessing step.) The proof is an easy induction on the number of moves of J and M. For the purposes of our proof, the configuration of a JTM will be described simply with the string $X_1 X_2 \ldots X_{i-1} q X_i X_{i+1} \ldots X_n$, where $X_1 X_2 \ldots X_n$ is the portion of the tape (lower track only) between the leftmost and rightmost non-blanks and $\overset{q}{X_i}$ is the symbol in the middle column of the scanning window (the entire portions of the upper track, on the left and right of q, are omitted). Consider the basis case, i.e. when $n = 0$ (M is yet to make a move, and J has just executed the zeroth move). M's initial configuration will be $q_0 X_1 X_2 \ldots X_n$, where $X_1 X_2 \ldots X_n = w$, the input string, and q_0, the initial state of M. J would have had $X_1 B X_2 \ldots X_n$ as its initial configuration (that is how we set it up), but, *after its zeroth move*, it will become $q_0 X_1 X_2 \ldots X_n$ (according to transition rule 1), as required. Now, let us suppose that the configuration of both M and J is $X_1 X_2 \ldots X_{i-1} q X_i X_{i+1} \ldots X_n$ after k moves. Then, after the $(k + 1)$th move, the configuration of *both* M and J will be either:

(i) $X_1 X_2 \ldots q' X_{i-1} Y X_{i+1} \ldots X_n$, if the transition rule for M is $\delta(q, X_i) = (q', Y, L)$
(J would have applied transition rule 2), or
(ii) $X_1 X_2 \ldots X_{i-1} Y q' X_{i+1} \ldots X_n$, if the transition rule for M is $\delta(q, X_i) = (q', Y, R)$
(J would have applied transition rule 3), as required. $\square$

4   Conclusion

It is traditionally known that Turing machines with only one state are not as powerful as their counterparts with more number of states. We have shown that this is no longer the case if we modify the TM a little: by adding an extra feature, namely, reading/writing a 3–letter string (as opposed to reading/writing a single symbol). While this result may turn out to be a product of mere theoretical curiosity, we believe that JTMs could serve as normal forms that might make proofs in computation theory simpler.

Acknowledgement

I thank my former colleague and friend, Mr. V. Ganesh Babu, for kindly reading a draft of my paper and for his remarks.

References

1. Hopcroft, J.E., Motwani, R., Ullman, J.D.: Introduction to Automata Theory, Languages, and Computation. Pearson Education, London (2006)
2. Turing, A.M.: On computable numbers with an application to the Entscheidungsproblem. Proc. London Math. Society 2(42), 230–265 (1936)

Polarizationless P Systems
with Active Membranes
Working in the Minimally Parallel Mode

Rudolf Freund[1], Gheorghe Păun[2], and Mario J. Pérez-Jiménez[3]

[1] Institute of Computer Languages
Vienna University of Technology, Favoritenstr. 9, Wien, Austria
rudi@emcc.at
[2] Institute of Mathematics of the Romanian Academy
PO Box 1-764, 014700 Bucureşti, Romania, and
Department of Computer Science and Artificial Intelligence
University of Sevilla, Avda. Reina Mercedes s/n, 41012 Sevilla, Spain
george.paun@imar.ro, gpaun@us.es
[3] Department of Computer Science and Artificial Intelligence
University of Sevilla, Avda. Reina Mercedes s/n, 41012 Sevilla, Spain
marper@us.es

Abstract. We investigate the computing power and the efficiency of P
systems with active membranes without polarizations, working in the
minimally parallel mode. Such systems are shown to be computationally
complete even when using only rules handling single objects in the membranes and avoiding the division of non-elementary membranes. Moreover, we elaborate an algorithm for solving **NP**-complete problems, yet
in this case we need evolution rules generating at least two objects as
well as rules for non-elementary membrane division.

1 Introduction

P systems with active membranes basically use five types of rules: (a) evolution
rules, by which a single object evolves to a multiset of objects, (b) send-in, and (c)
send-out rules, by which an object is introduced in or expelled from a membrane,
maybe modified during this operation into another object, (d) dissolution rules,
by which a membrane is dissolved, under the influence of an object, which may
be modified into another object by this operation, and (e) membrane division
rules; this last type of rules can be used both for elementary and non-elementary
membranes, or only for elementary membranes. As introduced in [9], all these
types of rules also use polarizations for membranes, "electrical charges" $+, -, 0$,
controlling the application of the rules.

Systems with rules of types (a), (b), (c) were shown to be equivalent in computational power with Turing machines [10], even when using only two polarizations [4], while P systems using all types of rules were shown to be universal even
without using polarizations [1]. Another important class of results concerns the

S.G. Akl et al.(Eds.): UC 2007, LNCS 4618, pp. 62–76, 2007.

possibility of using P systems with active membranes to provide polynomial solutions to computationally hard problems. Several papers have shown that both decision and numerical **NP**-complete problems can be solved in a polynomial time (often, even linear time) by means of P systems with three polarizations [10], then the number of polarizations was decreased to two [3]. The systems constructed for these solutions use only division of elementary membranes. At the price of using division also for non-elementary membranes, the polarizations can be removed completely [2].

All papers mentioned above apply the rules in the maximally parallel mode: in each computation step, the chosen multiset of rules cannot be extended anymore, i.e., no further rule could be added to this chosen multiset of rules in such a way that the resulting extended multiset still could be applied. Recently, another strategy of applying rules in parallel was introduced, the so-called minimal parallelism [5]: in each computation step, the chosen multiset of rules to be applied in parallel cannot be extended by any rule out of a set of rules from which no rule has been chosen so far for this multiset of rules in such a way that the resulting extended multiset still could be applied (this is not the only way to interpret the idea of minimal parallelism, e.g., see [6] for other possibilities, yet the results elaborated in this paper hold true for all the variants of minimal parallelism defined there). This introduces an additional degree of non-determinism in the system evolution, but still computational completeness and polynomial solutions to SAT were obtained in the new framework by using P systems with active membranes, with three polarizations and division of only elementary membranes.

In this paper we continue the study of P systems working in the minimally parallel way, and we prove that the polarizations can be avoided, at the price of using all five types of rules for computational completeness and even the division of non-elementary membranes for computational efficiency. On the other hand, in the proof for establishing computational completeness, in all types of rules we can restrict their form to handling only single objects in the membranes (we call this the one-normal form for P systems with active membranes).

2   Prerequisites

We suppose that the reader is familiar with the basic elements of Turing computability [7], and of membrane computing [10]. We here, in a rather informal way, introduce only the necessary notions and notation.

For an alphabet A, by A^* we denote the set of all strings of symbols from A including the empty string λ. A *multiset* over an alphabet A is a mapping from A to the set of natural numbers; we represent a multiset by a string from A^*, where the number of occurrences of a symbol $a \in A$ in a string w represents the multiplicity of a in the multiset represented by w. The family of Turing-computable sets of natural numbers is denoted by NRE.

In our proofs showing computational completeness we use the characterization of NRE by means of *register machines*. A *register machine* is a construct $M =$

(n, B, l_0, l_h, I), where n is the number of registers, B is a set of instruction labels, l_0 is the start label, l_h is the halt label (assigned to HALT only), and I is a set of instructions of the following forms:

- $l_i : (\text{ADD}(r), l_j, l_k)$ add 1 to register r, and then go to one of the instructions labeled by l_j and l_k, non-deterministically chosen;
- $l_i : (\text{SUB}(r), l_j, l_k)$ if register r is non-empty (non-zero), then subtract 1 from it and go to the instruction labeled by l_j, otherwise go to the instruction labeled by l_k;
- $l_h : \text{HALT}$ the halt instruction.

A register machine M generates a set $N(M)$ of numbers in the following way: start with the instruction labeled by l_0, with all registers being empty, and proceed to apply instructions as indicated by the labels and by the contents of registers. If we reach the halt instruction, then the number stored at that time in register 1 is taken into $N(M)$. It is known (e.g., see [8]) that in this way we can compute all the sets of numbers which are Turing-computable, even using register machines with only three registers as well as registers two and three being empty whenever the register machine halts.

3   P Systems with Active Membranes

A *P system with active membranes*, of the initial degree $n \geq 1$, is a construct of the form $\Pi = (O, H, \mu, w_1, \ldots, w_n, R, h_o)$ where

1. O is the alphabet of *objects*;
2. H is a finite set of *labels* for membranes;
3. μ is a *membrane structure*, consisting of n membranes having initially neutral polarizations, labeled (not necessarily in a one-to-one manner) with elements of H;
4. $w_1, \ldots, w_n$ are strings over O, describing the *multisets of objects* placed in the n initial regions of μ;
5. R is a finite set of *developmental rules*, of the following forms:
   (a) $[_h a \rightarrow v]_h^e$, for $h \in H, e \in \{+, -, 0\}, a \in O, v \in O^*$
       (*object evolution* rules);
   (b) $a[_h \,]_h^{e_1} \rightarrow [_h b]_h^{e_2}$, for $h \in H, e_1, e_2 \in \{+, -, 0\}, a, b \in O$
       (*in* communication rules);
   (c) $[_h a \,]_h^{e_1} \rightarrow b[_h \,]_h^{e_2}$, for $h \in H, e_1, e_2 \in \{+, -, 0\}, a, b \in O$
       (*out* communication rules);
   (d) $[_h a \,]_h^e \rightarrow b$, for $h \in H, e \in \{+, -, 0\}, a, b \in O$
       (*dissolving* rules);
   (e) $[_h a \,]_h^{e_1} \rightarrow [_h b \,]_h^{e_2}[_h c \,]_h^{e_3}$, for $h \in H, e_1, e_2, e_3 \in \{+, -, 0\}, a, b, c \in O$
       (*division* rules for elementary or non-elementary membranes in reaction with an object, being replaced in the two new membranes by possibly new objects; the remaining objects may have evolved in the same step by rules of type (a) and the result of this evolution is duplicated in the two new membranes);
6. $h_o \in H$ or $h_o = env$ indicates the output region.

In all types of rules considered above, the labels of membranes are never changed. All the rules of any type involving a membrane h constitute the set R_h, and for each occurrence of a membrane labelled by h we consider different copies of this set R_h. In the maximally parallel mode, in each step (a global clock is assumed) we apply multisets of rules in such a way that no further rule can be applied to the remaining objects or membranes. In each step, each object and each membrane can be involved in only one rule. In the minimally parallel mode, in each step, we choose a multiset of rules taken from the sets assigned to each membrane in the current configuration in such a way that after having assigned objects to all the rules in this multiset, no rule from any of the sets from which no rule has been taken so far, could be used in addition. In every mode, each membrane and each object can be involved in only one rule, and the choice of rules to be used and of objects and membranes to evolve is done in a non-deterministic way. For rules of type (a) the membrane is not considered to be involved: when applying $[_h a \rightarrow v]_h$, the object a cannot be used by other rules, but the membrane h can be used by any number of rules of type (a) as well as by one rule of types (b) – (e). In each step, the use of rules is done in the bottom-up manner (first the inner objects and membranes evolve, and the result is duplicated if any surrounding membrane is divided).

A halting computation provides a result given by the number of objects present in region h_o at the end of the computation; this is a region of the system if $h_o \in H$ (and in this case, for a computation to be successful, exactly one membrane with label h_o should be present in the halting configuration), or it is the environment if $h_o = env$.

A system Π is said to be in the *one-normal form* if the membranes have no polarization (it is the same as saying that always all membranes have the same polarization, say 0, which therefore is irrelevant and thus omitted) and the rules are of the forms $[_h a \rightarrow b]_h$, $a[_h \,]_h \rightarrow [_h b]_h$, $[_h a \,]_h \rightarrow b[_h \,]_h$, $[_h a \,]_h \rightarrow b$, and $[_h a \,]_h \rightarrow [_h b \,]_h [_h c \,]_h$, for $a, b, c \in O$, such that $a \neq b$, $a \neq c$, $b \neq c$.

The set of numbers generated in the minimally parallel way by a system Π is denoted by $N_{min}(\Pi)$. The family of sets $N_{min}(\Pi)$, generated by systems with rules of the non-restricted form, having initially at most n_1 membranes and using configurations with at most n_2 membranes during any computation is denoted by $N_{min}OP_{n_1,n_2}((a),(b),(c),(d),(e))$; when a type of rules is not used, it is not mentioned in the notation. If any of the parameters n_1, n_2 is not bounded, then it is replaced by $*$. If the systems do not use polarizations for membranes, then we write $(a_0),(b_0),(c_0),(d_0),(e_0)$ instead of $(a),(b),(c),(d),(e)$. When the system Π is in the one-normal form, then we write $N_{min}OP_{n_1,n_2}((a_1),(b_1),(c_1),(d_1),(e_1))$.

When considering families of numbers generated by P systems with active membranes working in the maximally parallel mode, the subscript min is replaced by max in the previous notations. For precise definitions, we refer to [10] and to the papers mentioned below.

4   Computational Completeness

The following results are well known:

Theorem 1. (i) $N_{max}OP_{3,3}((a),(b),(c)) = NRE$, [10].
$\quad\quad\quad\quad (ii)$ $N_{max}OP_{*,*}((a_0),(b_0),(c_0),(d_0),(e_0)) = NRE$, [1].
$\quad\quad\quad\quad (iii)$ $N_{min}OP_{3,3}((a),(b),(c)) = NRE$, [5].

In turn, the following inclusions follow directly from the definitions:

Lemma 1. $N_{mode}OP_{n_1,n_2}((a_1),(b_1),(c_1),(d_1),(e_1)) \subseteq$
$\quad\quad N_{mode}OP_{n_1,n_2}((a_0),(b_0),(c_0),(d_0),(e_0)) \subseteq$
$\quad\quad N_{mode}OP_{n_1,n_2}((a),(b),(c),(d),(e))$,
$\quad$ for all $n_1, n_2 \geq 1$, $mode \in \{max, min\}$.

We now improve the equalities from Theorem 1 in certain respects, starting with proving the computational completeness of P systems with active membranes in the one-normal form when working in the maximally parallel mode, and then we extend this result to the minimal parallelism.

Theorem 2. For all $n_1 \geq 5$,

$$N_{max}OP_{n_1,*}((a_1),(b_1),(c_1),(d_1),(e_1)) = NRE.$$

Proof. We only prove the inclusion

$$NRE \subseteq N_{max}OP_{5,*}((a_1),(b_1),(c_1),(d_1),(e_1)).$$

Let us consider a register machine $M = (3, B, l_0, l_h, I)$ generating an arbitrary set $N(M) \in NRE$. We construct the P system of initial degree 5

$$\Pi = (O, H, \mu, w_s, w_h, w_1, w_2, w_3, R, env),$$
$$O = \{d_i \mid 0 \leq i \leq 5\} \cup \{g, \#, \#', p, p', p'', c, c', c''\} \cup B \cup \{l' \mid l \in B\}$$
$$\quad\quad \cup \{l_{iu} \mid l_i \text{ is the label of an ADD instruction in } I, 1 \leq u \leq 4\}$$
$$\quad\quad \cup \{l_{iu0} \mid l_i \text{ is the label of a SUB instruction in } I, 1 \leq u \leq 4\}$$
$$\quad\quad \cup \{l_{iu+} \mid l_i \text{ is the label of a SUB instruction in } I, 1 \leq u \leq 6\},$$
$$H = \{s, h, 1, 2, 3\},$$
$$\mu = [_s[_1 \,]_1 \,[_2\,]_2 \,[_3 \,]_3[_h \,]_h]_s,$$
$$w_s = l_0 d_0, w_\alpha = \lambda, \text{ for all } \alpha \in H - \{s\},$$

with the rules from R constructed as described below.

The value stored in a register $r = 1, 2, 3$ of M, in Π is represented by the number of copies of membranes with label r plus one (if the value of the register r is k, then we have $k + 1$ membranes with label r). The membrane with label h is auxiliary, it is used for controlling the correct simulation of instructions of M by computations in Π. Each step of a computation in M, i.e., using an ADD or a SUB instruction, corresponds to six steps of a computation in Π. We start with all membranes being empty, except for the skin region, which contains the initial label l_0 of M and the auxiliary object d_0. If the computation in M halts, that is, the object l_h appears in the skin region of Π, then we pass to producing one object c for each membrane with label 1 present in the system, except one; in this way, the number of copies of c sent to the environment represents the correct result of the computation in M.

We now list the rules used in each of the six steps of simulating instructions ADD and SUB, and finally we present the rules for producing the result of the computation; in each configuration, only the membranes and the objects relevant for the simulation of the respective instruction are specified. In particular, we ignore the "garbage" object g, because once introduced it remains idle for the whole computation.

The **simulation of an instruction** $l_i : (\text{ADD}(r), l_j, l_k)$ uses the rules from Table 1. The label object l_i enters into the correct membrane r (even if the register r is empty, there is at least one membrane with label r) and in the next step divides it. The object l_{i2} exits the newly produced membrane, but g remains inside; l_{i2} will evolve three further steps, just to synchronize with the evolution of the auxiliary objects $d_u, u \geq 0$, and the auxiliary membrane h, so that in the sixth step we end up with the label l_j or l_k of the next instruction to be simulated present in the skin membrane, together with d_0; note that the number of copies of membrane r was increased by one.

Table 1. The simulation of an ADD instruction

Step	Main rules	Auxiliary rules	Configuration
–	–	–	$[_s l_i d_0 [_r]_r \cdots [_h]_h]_s$
1	$l_i[_r]_r \rightarrow [_r l_i']_r$	$[_s d_0 \rightarrow d_1]_s$	$[_s d_1 [_r l_i']_r \cdots [_h]_h]_s$
2	$[_r l_i']_r \rightarrow [_r l_{i1}]_r [_r g]_r$	$[_s d_1 \rightarrow d_2]_s$	$[_s d_2 [_r l_{i1}]_r [_r g]_r \cdots [_h]_h]_s$
3	$[_r l_{i1}]_r \rightarrow l_{i2}[_r]_r$	$d_2[_h]_h \rightarrow [_h d_3]_h$	$[_s l_{i2} [_r]_r [_r]_r \cdots [_h d_3]_h]_s$
4	$[_s l_{i2} \rightarrow l_{i3}]_s$	$[_h d_3]_h \rightarrow d_4[_h]_h$	$[_s l_{i3} d_4 [_r]_r [_r]_r \cdots [_h]_h]_s$
5	$[_s l_{i3} \rightarrow l_{i4}]_s$	$[_s d_4 \rightarrow d_5]_s$	$[_s l_{i4} d_5 [_r]_r [_r]_r \cdots [_h]_h]_s$
6	$[_s l_{i4} \rightarrow l_t]_s$, $t \in \{j, k\}$	$[_s d_5 \rightarrow d_0]_s$	$[_s l_t d_0 [_r]_r [_r]_r \cdots [_h]_h]_s$

The auxiliary objects $d_u, u \geq 0$, and the auxiliary membrane h are also used in the simulation of SUB instructions; in step 4, there also are other rules to be used in membrane h, introducing the trap object $\#$, but using such rules will make the computation never halt, hence, they will not produce an unwanted result.

The evolution of the auxiliary objects $d_u, u \geq 0$, is graphically represented in Figure 1; membrane h is only used in step 3, to send an object inside, and in the next step, for sending an object out of it. On each arrow we indicate the step, according to Table 1; the steps made by using rules of types (b) and (c) are pointed out by drawing the respective arrows crossing the membranes, thus suggesting the *in/out* actions.

The way the simulation of an ADD instruction works is depicted in Figure 2; the division operation from step 2 is indicated by a branching arrow.

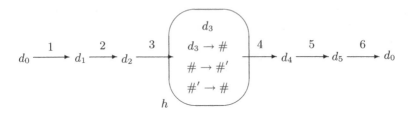

Fig. 1. The evolution of the auxiliary objects

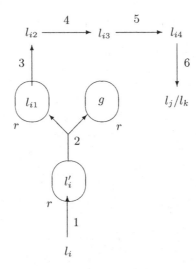

Fig. 2. The simlation of an ADD instruction

The **simulation of an instruction** $l_i : (\mathtt{SUB}(r), l_j, l_k)$ is done with the help of the auxiliary membrane h. The object l_i enters a membrane r (there is at least one copy of it) and divides it, and on this occasion makes a non-deterministic choice between trying to continue as having register r non-empty or as having it empty. If the guess has been correct, then the correct action is done (decrementing the register in the first case, doing nothing in the second case) and the correct next label is introduced, i.e., l_j or l_k, respectively. If the guess has not been correct, then the trap object $\#$ is introduced. For all membranes x of the system we consider the rules $[_x\# \to \#']_x$, $[_x\#' \to \#]_x$, hence, the appearance of $\#$ will make the computation last forever.

In Table 2 we present the rules used in the case of guessing that the register r is not empty. Like in the case of simulating an ADD instruction, the auxiliary objects and membranes do not play any role in this case. The evolution of the objects l_{iu+} in the six steps described in Table 2 is depicted in Figure 3; when a dissolving operation is used, this is indicated by writing δ near the respective arrow.

In step 2 we divide the membrane r containing the object l_i' and the objects l_{i1+}, l_{i2+} are introduced in the two new membranes. One of them is immediately

Table 2. The simulation of a SUB instruction, guessing that register r is not empty

Step	Main rules	Auxiliary rules	Configuration
–	–	–	$[_s l_i d_0 [_r]_r \cdots [_h]_h]_s$
1	$l_i[_r]_r \to [_r l_i']_r$	$[_s d_0 \to d_1]_s$	$[_s d_1 [_r l_i']_r \cdots [_h]_h]_s$
2	$[_r l_i']_r \to$ $[_r l_{i1+}]_r [_r l_{i2+}]_r$	$[_s d_1 \to d_2]_s$	$[_s d_2 [_r l_{i1+}]_r [_r l_{i2+}]_r$ $\cdots [_h]_h]_s$
3	$[_r l_{i1+}]_r \to l_{i3+}$ $[_r l_{i2+}]_r \to l_{i4+}[_r]_r$	$d_2[_h]_h \to [_h d_3]_h$	$[_s l_{i3+} l_{i4+}[_r]_r \cdots [_h d_3]_h]_s$
4	$l_{i3+}[_r]_r \to [_r g]_r$ $l_{i4+}[_r]_r \to [_r l_{i5+}]_r$ or $[_s l_{i3+} \to \#]_s$, $[_s l_{i4+} \to \#]_s$	$[_h d_3]_h \to d_4[_h]_h$	$[_s d_4[_r l_{i5+}]_r [_r g]_r \cdots [_h]_h]_s$ $[_s d_4 \#[_r]_r \cdots [_h]_h]_s$
5	$[_r l_{i5+}]_r \to l_{i6+}$	$[_s d_4 \to d_5]_s$	$[_s d_5 l_{i6+}[_r]_r \cdots [_h]_h]_s$
6	$[_s l_{i6+} \to l_j]_s$	$[_s d_5 \to d_0]_s$	$[_s l_j d_0[_r]_r \cdots [_h]_h]_s$

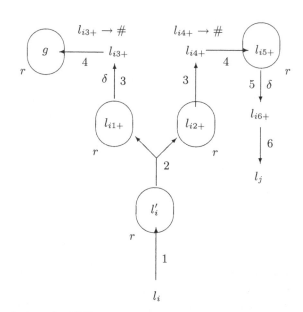

Fig. 3. The simulation of a SUB instruction, guessing that the register is non-empty

dissolved, thus the number of copies of membrane r remains unchanged; the objects l_{i3+}, l_{i4+} are introduced in this step. In the next step (the fourth one of the simulation), objects l_{i3+}, l_{i4+} look for membranes r in the skin region. If both of them find such membranes – and this is the correct/desired continuation – then both of them enter such membranes; l_{i3+} becomes g and l_{i4+} becomes l_{i5+}. If only one of them finds a membrane r, then the other one has to evolve to object $\#$ in the skin membrane and the computation never halts.

If we had enough membranes r and l_{i3+}, l_{i4+} went there, then in the next step l_{i5+} dissolves one membrane r thereby being changed to l_{i6+}; in this way, the

Table 3. The simulation of a SUB instruction, guessing that register r is empty

Step	Main rules	Auxiliary rules	Configuration
–	–	–	$[_s l_i d_0 [_r]_r \cdots [_h]_h]_s$
1	$l_i[_r]_r \to [_r l_i']_r$	$[_s d_0 \to d_1]_s$	$[_s d_1 [_r l_i']_r \cdots [_h]_h]_s$
2	$[_r l_i']_r \to$ $[_r l_{i10}]_r[_r l_{i20}]_r$	$[_s d_1 \to d_2]_s$	$[_s d_2 [_r l_{i10}]_r[_r l_{i20}]_r$ $\cdots [_h]_h]_s$
3	$[_r l_{i10}]_r \to l_{i30}$ $[_r l_{i20}]_r \to l_{i40}[_r]_r$	$d_2[_h]_h \to [_h d_3]_h$	$[_s l_{i30} l_{i40} [_r]_r$ $\cdots [_h d_3]_h]_s$
4	$l_{i30}[_r]_r \to [_r g]_r$, l_{i40} waits or $l_{i40}[_r]_r \to [_r \#]_r$ or $l_{i40}[_h]_h \to [_h l_k']_h$	$[_h d_3]_h \to d_4[_h]_h$ $[_h d_3 \to \#]_h$	$[_s l_{i40} d_4[_r g]_r \cdots [_h]_h]_s$ $[_s d_4[_r \#]_r \cdots [_h]_h]_s$ $[_s [_r g]_r \cdots [_h l_k'\#]_h]_s$
5	$l_{i40}[_h]_h \to [_h l_k']_h$	$[_s d_4 \to d_5]_s$	$[_s d_5[_r]_r \cdots [_h l_k']_h]_s$
6	$[_h l_k']_h \to l_k[_h]_h$	$[_s d_5 \to d_0]_s$	$[_s l_k d_0[_r]_r \cdots [_h]_h]_s$

number of membranes r has successfully been decremented. In the sixth step, l_{i6+} finally becomes l_j (simultaneously with regaining d_0), and this completes the simulation.

However, in step 2, instead of the rule $[_r l_i']_r \to [_r l_{i1+}]_r[_r l_{i2+}]_r$ we can use the rule $[_r l_i']_r \to [_r l_{i10}]_r[_r l_{i20}]_r$, with the intention to simulate the subtract instruction in the case when the register r is empty – that is, only one membrane with label r is present in the system. The rules used in the six steps of the simulation are given in Table 3.

The six steps of the simulation of a SUB instruction when guessing the corresponding register to be empty are shown in Figure 4, this time with the evolution of the auxiliary objects being indicated, too. The arrows marked with $\#$ correspond to computations which are not desired.

In this case, the auxiliary objects $d_u, u \geq 0$, and the auxiliary membrane h play an essential role. Until step 3 we proceed exactly as above, but now we introduce the objects l_{i30} and l_{i40}. In step 4, l_{i30} then can enter the available membrane r, there evolving to g. In this fourth step, there is a competition between l_{i40} and d_3 for membrane h, but if l_{i40} enters membrane h already in this step, then the trap object $\#$ is produced here by the rule $[_h d_3 \to \#]_h$.

If there is a second membrane r, i.e., if the guess has been incorrect, then – to avoid the scenario described before – the rule $l_{i40}[_r]_r \to [_r \#]_r$ has to be used simultaneously with $l_{i30}[_r]_r \to [_r g]_r$ and $[_h d_3]_h \to d_4[_h]_h$ in the fourth step, yet then the computation never ends, too. If there is no second membrane r, then, in step 4, l_{i40} has to wait unchanged in the skin region if we apply $l_{i30}[_r]_r \to [_r g]_r$ and $[_h d_3]_h \to d_4[_h]_h$ in the desired way. Thus, the only way not to introduce the object $\#$ is (i) not to have a second membrane r, (ii) to use the rule $[_h d_3]_h \to d_4[_h]_h$, thus preventing the use of the rule $l_{i40}[_h]_h \to [_h l_k']_h$ already in the fourth step, and (iii) not sending l_{i40} to the unique membrane r. After having waited unchanged in the skin region during step 4, the object l_{i40} in the next step should enter membrane h (if entering the unique membrane r,

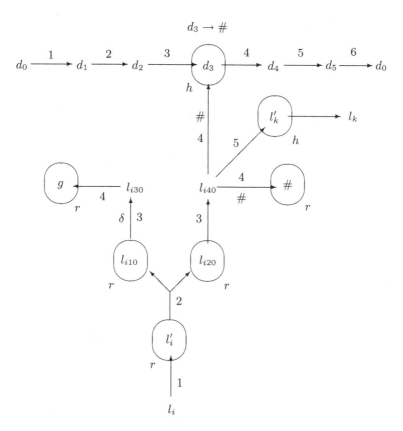

Fig. 4. The simulation of a SUB instruction, guessing that the register is empty

it becomes # there), and finally, l_k is released from membrane h, at the same time when d_0 appears again.

If the computation in M halts by reaching l_h, i.e., if the object l_h is introduced in the skin region, we start to use the following rules:

$$l_h[_1\ ]_1 \to [_1 l'_h]_1, [_1 l'_h]_1 \to p, p[_1\ ]_1 \to [_1 p']_1,$$
$$[_1 p']_1 \to [_1 p'']_1 [_1 c']_1, [_1 p'']_1 \to p, [_1 c']_1 \to c'', [_s c'']_s \to c[_s\ ]_s,$$
$$p[_h\ ]_h \to [_h p']_h, [_h p']_h \to p.$$

The object l_h dissolves one membrane with label 1 (thus, the remaining membranes with this label are now as many as the value of register 1 of M), and gets transformed into p. This object enters each membrane with label 1, divides it and then reproduces itself by passing through p' and p'', thereby also dissolving one membrane with label 1, and also produces a copy of the object c'', which immediately exits the system, thereby being changed into c.

At any time, the object p can also enter membrane h and then dissolve it. The computation can stop only after all membranes with label 1 having been dissolved (i.e., a corresponding number of copies of c has been sent out) and

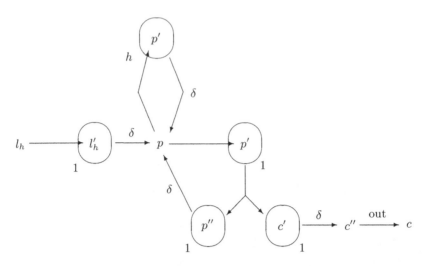

Fig. 5. The end of computations

after the single membrane h having been dissolved, too, i.e., the evolution of objects d_u, $u \geq 0$, is also stopped. The function of this final module is described in Figure 5.

In sum, we conclude $N(M) = N_{max}(\Pi)$. $\square$

Theorem 3 gives the same result as Theorem 3 from [1], with the additional constraint to have evolution rules with exactly one object on the right-hand side (neither producing more objects, nor erasing objects, as it is the case in [1]).

The previous proof can easily be changed in order to obtain the computational completeness of P systems in the one-normal form also in the case of minimal parallelism. Specifically, we can introduce additional membranes in order to avoid having two or more evolution rules to be applied in the same region or one or more evolution rules and one rule of types (b) – (e) which involve the same membrane:

- Applying any of the rules $[_s l_{i3+} \rightarrow \#]_s$ and $[_s l_{i4+} \rightarrow \#]_s$, in step 4 of the simulation of SUB for the guess of a non-empty register, means introducing the trap object and is the same for the fate of the computation as applying all possible rules of that kind.
- Steps 5 and 6 for the auxiliary rules are performed in the skin region, in parallel with steps of the main rules in the simulation of incrementing or decrementing a register. This can be avoided by introducing one further auxiliary membrane, h', in the skin region, and replacing the rules $[_s d_4 \rightarrow d_5]_s$, $[_s d_5 \rightarrow d_0]_s$ (both of them from R_s) by the rules $d_4[_{h'}]_{h'} \rightarrow [_{h'} d_5]_{h'}$ and $[_{h'} d_5]_{h'} \rightarrow d_0[_{h'}]_{h'}$. These rules are associated with membrane h', hence, they should be used in parallel with the rules from the skin region.
- A further case when the maximal parallelism is important in the previous construction is in step 4 of simulating a SUB instruction for the guess that

the register is empty: if rule $l_{i40}[_h \ ]_h \to [_h l'_k]_h$ is used in step 4, then also the rule $[_h d_3 \to \#]_h$ must be used. In the minimally parallel mode, this can be ensured if we replace this latter rule by $d_3[_{h''} \ ]_{h''} \to [_{h''} \#]_{h''}$, where h'' is one further membrane, provided in the initial configuration inside membrane h. In this way, the rule $d_3[_{h''} \ ]_{h''} \to [_{h''} \#]_{h''}$ involves the new membrane h'', hence, it should be used in parallel with the rule $l_{i40}[_h \ ]_h \to [_h l'_k]_h$. Of course, we also need the rules $[_{h''} \# \to \#']_{h''}$, $[_{h''} \#' \to \#]_{h''}$, instead of the corresponding rules from membrane h.

In this way, we obtain the following counterpart of Theorem 2 (note that we use two further membranes, h' and h''):

Theorem 3. *For all $n_1 \geq 7$,*

$$N_{min}OP_{n_1,*}((a_{s1}),(b_{s1}),(c_{s1}),(d_{s1}),(e_{s1})) = NRE.$$

The previous systems can easily be modified in order to work in the accepting mode: we introduce the number to be analyzed in the form of the multiplicity of membranes with label 1 initially present in the skin membrane (for number n we start with $n + 1$ membranes $[_1 \ ]_1$ in the system) and we work exactly as above, with the final module now having only the task of dissolving the auxiliary membrane with label h, thus halting the computation. Consequently, all the previous results, for both minimally and maximally parallel use of rules, are valid also for accepting P systems in the one-normal form.

5   Solving SAT by Polarizationless P Systems

We now turn our attention to applying P systems for deciding **NP**-complete problems and give an efficient semi–uniform solution to the satisfiability problem by using P systems with polarizationless active membranes working in the minimally parallel mode.

Theorem 4. *The satisfiability of any propositional formula in the conjunctive normal form, using n variables and m clauses, can be decided in a linear time with respect to n by a polarizationless P system with active membranes, constructed in linear time with respect to n and m, and working in the minimally parallel mode.*

Proof. Let us consider a propositional formula $\varphi = C_1 \wedge \ldots \wedge C_m$ such that each clause C_j, $1 \leq j \leq m$, is of the form $C_j = y_{j,1} \vee \ldots \vee y_{j,k_j}$, $k_j \geq 1$, for $y_{j,r} \in \{x_i, \neg x_i \mid 1 \leq i \leq n\}$. For each $i = 1, 2, \ldots, n$, let us consider the two substrings $t(x_i)$ and $f(x_i)$ of $c_1 \ldots c_m$ such that, with $1 \leq j \leq m$,

$t(x_i)$ exactly contains the c_j such that $y_{j,r} = x_i$ for some r, $1 \leq r \leq k_j$,

$f(x_i)$ exactly contains the c_j such that $y_{j,r} = \neg x_i$ for some r, $1 \leq r \leq k_j$.

Thus, $t(x_i)$ and $f(x_i)$ describe the sets of clauses which take the value *true* when x_i is *true* and when x_i is *false*, respectively.

We construct the P system $\Pi(\varphi)$ with the following components (the output membrane is not necessary, because the result is obtained in the environment):

$$O = \{a_i, f_i, t_i \mid 1 \leq i \leq n\} \cup \{c_j, d_j \mid 1 \leq j \leq m\} \cup \{p_i \mid 1 \leq i \leq 2n+7\}$$
$$\cup \{q_i \mid 1 \leq i \leq 2n+1\} \cup \{r_i \mid 1 \leq i \leq 2n+5\} \cup \{b_1, b_2, y, \text{yes}, \text{no}\},$$
$$H = \{s, s', p, q, r, 0, 1, 2, \ldots, m\},$$
$$\mu = [_s[_{s'}[_p \ ]_p[_0[_q \ ]_q[_r]_r[_1 \ ]_1[_2 \ ]_2 \cdots [_m \ ]_m]_0]_{s'}]_s,$$
$$w_p = p_1, \ w_q = q_1, \ w_r = r_1, \ w_0 = a_1,$$
$$w_s = w_{s'} = \lambda, \ w_j = \lambda, \ \text{for all } j = 1, 2, \ldots, m.$$

The set of evolution rules, R, consists of the following rules:

(1) $[p_i \rightarrow p_{i+1}]_p$, for all $1 \leq i \leq 2n+6$,
   $[q_i \rightarrow q_{i+1}]_q$, for all $1 \leq i \leq 2n$,
   $[r_i \rightarrow r_{i+1}]_r$, for all $1 \leq i \leq 2n+4$.
   These rules are used for evolving counters p_i, q_i, and r_i in membranes with labels p, q, and r, respectively.
(2) $[a_i]_0 \rightarrow [f_i]_0[t_i]_0$, for all $1 \leq i \leq n$,
   $[f_i \rightarrow f(x_i)a_{i+1}]_0$ and $[t_i \rightarrow t(x_i)a_{i+1}]_0$, for all $1 \leq i \leq n-1$,
   $[f_n \rightarrow f(x_n)]_0$, $[t_n \rightarrow t(x_n)]_0$.
   The goal of these rules is to generate the truth assignments of the n variables $x_1, \ldots, x_n$, and to analyze the clauses satisfied by x_i and $\neg x_i$, respectively.
(3) $c_j[\ ]_j \rightarrow [c_j]_j$ and $[c_j]_j \rightarrow d_j$, for all $1 \leq j \leq m$.
   In parallel with the division steps, if a clause C_j is satisfied by the previously expanded variable, then the corresponding object c_j enters membrane j in order to dissolve it and to send objects d_j to membrane 0.
(4) $[q_{2n+1}]_q \rightarrow q_{2n+1}[\ ]_q$, $[q_{2n+1} \rightarrow b_1]_0$.
   The counter q produces an object b_1 in each membrane 0.
(5) $b_1[\ ]_j \rightarrow [b_1]_j$ and $[b_1]_j \rightarrow b_2$, for all $1 \leq j \leq m$, $[b_2]_0 \rightarrow b_2$.
   These rules allow to detect whether the truth assignment associated with a membrane 0 assigns the value *false* to the formula (in that case, the membrane 0 will be dissolved).
(6) $[p_{2n+7}]_p \rightarrow p_{2n+7}[\ ]_p$, $[p_{2n+7}]_{s'} \rightarrow \text{no}[\ ]_{s'}$, $[\text{no}]_s \rightarrow \text{no}[\ ]_s$,
   $[r_{2n+5}]_r \rightarrow r_{2n+5}$, $[r_{2n+5}]_0 \rightarrow y[\ ]_0$, $[y]_{s'} \rightarrow \text{yes}$, $[\text{yes}]_s \rightarrow \text{yes}[\ ]_s$.
   These rules produce the answer of the P system.

The membranes with labels p, q, and r, with the corresponding objects p_i, q_i, and r_i, respectively, are used as counters, which evolve simultaneously with the *main membrane* 0, where the truth assignments of the n variables $x_1, \ldots, x_n$ are generated; the use of separate membranes for counters makes possible the correct synchronization even for the case of the minimal parallelism. The evolution of counters is done by the rules of type (1). In parallel with these rules, membrane 0 evolves by means of the rules of type (2). In odd steps (from step 1 to step $2n$), we divide the (non-elementary) membrane 0 (with f_i, t_i corresponding to the truth values *false, true*, respectively, for variable x_i); in even steps we introduce the clauses satisfied by $x_i, \neg x_i$, respectively. After $2n$ steps, all 2^n truth assignments for the n variables have been generated and they are encoded in membranes labeled by 0. If a clause C_j is satisfied by the previously expanded variable, then the corresponding object c_j enters membrane j, by means of the first rule of type (3), permitting its dissolution by means of the second rule of that type and sending objects d_j to membrane 0. This is done also in step $2n + 1$, in parallel with using the rules of type (1) and (4) for evolving membranes p, q, and r. In step $2n + 2$, the counters p_i and r_i keep evolving and the second rule of type (4) produces an object b_1 in each membrane 0.

After $2n + 2$ steps, the configuration $\mathcal{C}_{2n+2}$ of the system consists of 2^n copies of membrane 0, each of them containing the membrane q empty, the membrane r with the object r_{2n+3}, possible objects c_j and d_j, $1 \leq j \leq m$, as well as copies of those membranes with labels $1, 2, \ldots, m$ corresponding to clauses which were not satisfied by the truth assignment generated in that copy of membrane 0. The membranes associated with clauses satisfied by the truth assignments generated have been dissolved by the corresponding object c_j. The membrane p contains the object p_{2n+3}, and membrane regions s' and s are empty. Therefore, formula φ is satisfied if and only if there is a membrane 0 where all membranes $1, 2, \ldots, m$ have been dissolved. The remaining rules allow for sending out an object **yes** to the environment in step $2n + 8$, and, if the formula is not satisfiable, then the object **no** appears in step $2n + 9$. In both cases, this is the last step of the computation.

The system $\Pi(\varphi)$ uses $9n + 2m + 18$ objects, $m + 6$ initial membranes, containing in total 4 objects, and $9n + 4m + 20$ rules. The number of objects generated by any object evolution rule is bounded by $m + 1$. Clearly, all computations stop (after at most $2n + 9$ steps) and all give the same answer, **yes** or **no**, to the question whether formula φ is satisfiable, hence, the system is weakly confluent. These observations conclude the proof.   □

We can also design a *deterministic* P system $\Pi(\varphi)$ working in minimally parallel mode which decides the satisfiability of φ; to this aim, it is enough to have m copies of the object b_1 in each membrane 0 of the configuration $\mathcal{C}_{2n+2}$, which can be obtained by replacing the rule $[\, q_{2n+1} \rightarrow b_1 \,]_0$ by the rule $[\, q_{2n+1} \rightarrow b_1^m \,]_0$.

The system used in the proof of Theorem 4 is not in the one-normal form, because some of the rules of type (b) in (2) are of the form $[\,_h a \rightarrow u\,]_h$ with u being an arbitrary multiset, hence, it remains an interesting open problem to find a polynomial solution to **SAT** by a system in the one-normal form.

Acknowledgements. The work of Gh. Păun was partially supported by Project BioMAT 2-CEx06-11-97/19.09.06. The work of M.J. Pérez-Jiménez was supported by the project TIN2005-09345-C04-01 of the Ministerio de Educación y Ciencia of Spain, co–financed by FEDER funds, and of the project of Excellence TIC 581 of the Junta de Andalucía. The authors also gratefully acknowledge the useful comments of an unknown referee.

References

1. Alhazov, A.: P systems without multiplicities of symbol-objects. Information Processing Letters 100, 124–129 (2006)
2. Alhazov, A., Pérez-Jiménez, M.J.: Uniform Solution to QSAT Using Polarizationless Active Membranes. In: Vol. I of Proc. of the Fourth Brainstorming Week on Membrane Computing, Sevilla (Spain), pp. 29–40 (January 30 - February 3, 2006)
3. Alhazov, A., Freund, R.: Membrane Computing, International Workshop, WMC5, Milano, Italy, 2004, Selected Papers. In: Mauri, G., Păun, G., Pérez-Jiménez, M.J., Rozenberg, G., Salomaa, A. (eds.) WMC 2004. LNCS, vol. 3365, pp. 81–94. Springer, Heidelberg (2005)
4. Alhazov, A., Freund, R., Păun, Gh.: Computational completeness of P systems with active membranes and two polarizations. In: Margenstern, M. (ed.) MCU 2004. LNCS, vol. 3354, pp. 82–92. Springer, Heidelberg (2005)
5. Ciobanu, G., Pan, L., Păun, Gh., Pérez-Jiménez, M.J.: P systems with minimal parallelism. Theoretical Computer Science (to appear)
6. Freund, R., Verlan, S.: A formal framework for P systems. In: Proceedings WMC 8, 2007 (to appear)
7. Hopcroft, J.E., Ullman, J.D.: Introduction to Automata Theory, Languages, and Computation. Addison-Wesley, Reading, MA (1979)
8. Minsky, M.: Computation – Finite and Infinite Machines. Prentice-Hall, Englewood Cliffs, NJ (1967)
9. Păun, Gh.: P systems with active membranes: Attacking NP-complete problems. Journal of Automata, Languages and Combinatorics 6(1), 75–90 (2001)
10. Păun, Gh.: Membrane Computing. An Introduction. Springer, Heidelberg (2002)

On One Unconventional Framework for Computation

Lev Goldfarb

Faculty of Computer Science
University of New Brunswick
Fredericton, NB
goldfarb@unb.ca
http://www.cs.unb.ca/~goldfarb

Abstract. My main objective is to point out a *fundamental* weakness in the conventional conception of computation and suggest a way out. This weakness is directly related to a gross underestimation of the role of object representation in a computational model, hence confining such models to an unrealistic (input) environment, which, in turn, lead to "unrealistic" computational models. This lack of appreciation of the role of *structural object representation* has been inherited from logic and partly from mathematics, where, in the latter, the centuries-old tradition is to represent objects as unstructured "points". I also discuss why the appropriate fundamental reorientation in the conception of computational models will bring the resulting study of computation closer to the "natural" computational constrains. An example of the pertinent, class-oriented, representational formalism developed by our group over many years—Evolving Transformation System (ETS)—is briefly outlined here, and several general lines of research are suggested.

1 Introduction

Historically, computability theory has emerged within logic[1], and so it is not surprising that "logical" agenda continues to play an important part in its development[2]. This origin has determined to a considerable extent the basic "logical orientation" of the field, i.e. the kinds of questions that one might be asking (computable vs. non-computable, etc.).[3]

---

[1] E.g. Alan Turing proposed the Turing machine as a convenient mechanism for answering question of whether satisfiability in the predicate calculus was a solvable problem or not.

[2] E.g. Stephen Cook [1]: "When I submitted the abstract for this paper in December 1970 my interest was in predicate calculus theorem proving procedures, and I had not yet thought of the idea of NP completeness."

[3] In connection with this orientation of the field, it is useful to recall ideas of John Von Neumann, one of the leading prewar logicians who was well aware of the original Turing work and actually invited Turing to stay at the Institute for Advanced Studies as his assistant (after Turing completed his Princeton dissertation): "We are very

S.G. Akl et al.(Eds.): UC 2007, LNCS 4618, pp. 77–90, 2007.
© Springer-Verlag Berlin Heidelberg 2007

At the same time, already in his original paper, Turing had to face the issue of representation, both the "data" representation (on the tape), i.e. the input "symbol space"[4], and the representation of the machine itself (machine instructions and its states).

Following the original Turing's considerations, related to "the process of [symbol] recognition", it is not difficult to see that if these "symbols" would be endowed with more general structure—as is the case with all real-world "symbols"—the structure of the corresponding "generalized Turing machine", or "intelligent machine", would have to be fundamentally modified in order to be able to deal with such structured symbols. So, first of all, we are faced with the ubiquitous issue of *structural representation*. Again, I would like to emphasize that, when dealing with natural objects, the treatment of the issues related to the structural object representation must precede the computational considerations, since the formal structure of the latter should depend on the formal structure of the former: the structure of object operations will determine the structure of the corresponding basic "machine" operations. Perhaps an analogy may help. If we recall the concept of a mathematical structure (e.g. group, vector space, topological space, partially ordered set), we should favor the idea that the *structure of the machine* operating on the *data that is viewed as an element of a particular (fixed) mathematical structure* should depend on the underlying operations postulated within this mathematical structure. In other words, representational considerations should precede computational ones.

However simple the above considerations may look, the looks are deceiving: it turns out that so far mathematics (and logic) has not adequately addressed the issue of structural object representation, for good reasons. I suggest that these reasons are quite profound and have to do with the context within which the development of structural representation should take place. This context, of *classes and induction*, has never been adequately delineated neither within

---

far from possessing a theory of automata which deserves that name, that is, a purely mathematical-logical theory. There exists today a very elaborate system of formal logic and specifically, of logic as applied to mathematics. ... About the inadequacies [of logic], however, this may be said: Everybody who has worked in formal logic will confirm that it is one of the technically most refractory parts of mathematics. The reason for this is that it deals with rigid, all-or-none concepts .... The theory of automata, of the digital, all-or-none type ... is certainly a chapter in formal logic. It would, therefore, seem that it will have to share this unattractive property of formal logic." [2].

[4] "If we were to allow an infinity of symbols, then there would be symbols differing to an arbitrary small extent. ... The new observed squares must be immediately recognizable by the computer. ... Now if these squares are marked only by single symbols there can be only a finite number of them, and we should not upset our theory by adjoining these marked squares to the observed squares. If, on the other hand, ... [each of them is] marked by a sequence of symbols [i.e. by a *structured* symbol], we cannot regard *the process of recognition* as a simple process. This is a fundamental point and should be illustrated. [He then considers symbols formed by numeric sequences.]" [3, my emphasis]

philosophy—in spite of continuous attention it attracted over many centuries—nor within psychology, artificial intelligence, and information retrieval.

This paper is motivated by a recent development of the first such formalism for structural representation, the Evolving Transformation System (ETS). The reason I used adjective "first" has to do with the following considerations: despite the common usage, strings and graphs cannot be considered as satisfactory forms of structural representation. Briefly, a string, for example, does not carry *within itself* enough representational information to allow for the inductive recovery of the corresponding grammar, which, as it turns out, is a serious indication of the inadequacy of the underlying representational formalism. Not surprisingly, the development of such a radically different formalism as ETS is still in its initial stages, and hence, in this paper, it is prudent to focus on the overall directions of its development rather than on any concrete formal results.

I should also mention that the above reorientation in the conception of a computational model is expected to bring it closer to a (structural) generalization of the conventional mathematical structures, in which the basic formal/underlying structure is postulated axiomatically.

2   On a Proper Approach to the Concept of Structural Object Representation

Historically, mathematics has been the main, if not the only, "science of representation", so it is not surprising that the lack of appreciation of the role of object representation has been inherited from mathematics, where the centuries-old tradition has been to represent objects as unstructured "points". In this sense, the underestimation of the role of object representation in a computational model is also not surprising. But while the mainstream mathematics—having been isolated from the problems arising in computer science—had no *apparent* impetus to proceed with the necessary radical developments, computer science has no such excuse. The concepts of data structure and abstract data type have been "screaming" for such fundamental developments. As was mentioned in Introduction, I believe the reasons for the lack of such developments within computer science have to do with its historical roots in mathematical logic, whose development was not at all motivated[5] by the representation of "physical" objects [4], [5].

First, I propose to proceed with the development of formalism for structural representation via generalization of the most basic entities on which the entire edifice of mathematics and natural sciences stands, i.e. via structural generalization of the natural numbers: at this stage, I don't think we have a more fundamental choice for our starting point. I do not have in mind here relatively "minor" (historical) generalizations such as complex numbers, quaternions, etc., since such generalizations are still numeric-based. What I have in mind is more

---

[5] I.e. considerably less so than is the case with mathematics, which always *had to address* the concerns of physics.

far-reaching generalization, based on the generalization of the Peano constructive process in which a single simple successor operation is replaced by several
structural ones. Thus I am suggesting entities that are structured in a manner
shown in Figs. 1, 2.

Second, I suggest that the notion of structural representation can properly
be addressed only within the context of a class. The reasons have to do with
the following hypothesized ontology of objects. As we know, objects in nature
do not pop up out of nowhere but always take time to appear, and in each case
the way an object "appears" is similar to the way some other, "similar", objects
appear, i.e. an object always appears as an element of some *class* of closely
related objects, be it an atom, a stone, a protein, or a stop sign. Moreover, an
object co-evolves also together with its class: there is an "invisible" permanent
"bond" between an object and its class.

The crux of the proposed informational view of the universe can be expressed
in the form of the *first ontological postulate*: in addition to the classes themselves, for each class there exists, in some form, (its) *class representation*, which
maintains the integrity of the class by guiding the generation of its new elements
(and which therefore evolves together with the class). One should keep in mind
that this postulate can be seen as a continuation of a very remarkable line of
thought going back to Aristotle, Duns Scotus, and Francis Bacon, among others.

The *second ontological postulate* deals with the closely interconnected issue
related to the nature of object representation: as is the case with a biological organism, any object in nature exists along with its formative/generative history,
which is recoverable from the above class representation. Hence the "similarity" of objects should now be understood as the "similarity" of their formative
histories. Since it is the object's formative history that reveals its similarity or
dissimilarity with other objects, this formative history must be captured in the
object's "representation" (which is, in a sense, a generalization the fact that all
biological organisms store their developmental information). Actually, it is not
difficult to see that the two postulates are closely related, so that one "leads" to
the other.

Consistent with the above two postulates, one should approach the task of
developing a formalism for structural representation as that of developing a
class-oriented representational formalism[6], which would automatically ensure
that the formalism will be suitable for the purposes of inductive learning—since
the object representation would carry much more information about its class
representation—and consequently for the purposes of AI in general.

It is useful to emphasize that the development of such an informational formalism should not wait for the verification of the above hypotheses: for one
thing, no existing informational formalism insisted on any underlying hypothesis about the informational structure of the universe. I believe, however, that
in contrast to such practices, it is important to state up front an informational
hypothesis which would clarify and inform the development of the corresponding
informational formalism.

---

[6] My papers [4], [5] are recommended as clarifying related issues.

In what follows, I will proceed under the assumption that the above two postulates are adopted, at least as far as the development of formalism for structural
representation is concerned. Again, I would like to emphasize the importance of
such adoption: without the appropriate guiding considerations, the very notion
of object representation becomes far too ambiguous, as can be seen from its
development so far. For example, strings and graphs are not appropriate models
for structural representation, since, as mentioned above, neither a string nor a
graph carry within itself sufficient information for the inductive recovery of the
class *from which it is supposed to come*. This is, in part, a consequence of the
situation when none of the above two postulates have been adopted.

One more point regarding these postulates. If we recall Georg Cantor's conception of the set as "the multitude that might be thought as oneness", we can
see that the most natural way to conceive the multitude as one is to view these
objects as elements of one class (specified via some generative mechanism, see
section 4). Thus, within the proposed "computational" setting, the underlying
formal structures come from those of classes.

3   The Price of Relying on Conventional Discrete Representations

In this section, I want to address briefly the issue why the representation should
be of major concern when developing a computational model. First of all, I
assume (correctly or incorrectly) that the main objective of the present series of
conferences is to converge on *the concept of computation* that would be viable not
just within the "classical", i.e. "logical", setting but also in the natural sciences
in general.

The main problem with the conventional representations, such as strings,
graphs, etc., is that they do not incorporate within themselves any particular
(generative) structure that might be associated with their formation. However, as
postulated above (see the two postulates in section 2), this generative/formative
structure must be an integral part of an object's identity, and when absent creates
an unrealistic and ambiguous situation as to this identity. Such *representational*
deficiency is obviously impossible to overcome. Indeed, for a string **abbaca**,
there is simply no way to know the sequence of operations that was responsible for the string formation and, potentially, there are exponentially many of
them. As a consequence, from the very beginning we are critically handicapped
in regards to our ability to discover the underlying generative structure, and
no computational setting would be able to overcome this representational deficiency. To repeat, no amount of algorithmic maneuvering can recover the missing
representational information.

One might think that there is a simple way out of this situation: embed the
necessary formative information (via some tags) into the representation, end of
the story. However, this is a typical representational kludge, as opposed to a
scientific approach: we have not gotten wiser neither about structural object
representation, nor about the concept of "formative history".

So now we come to a considerably less trivial question: What is a general formalism for structural representation that would also explicate the notion of *structural generativity* as its integral part? Note that conventional computational formalisms, for example Chomsky's generative grammars, have not addressed this issue simply because they had implicitly assumed that strings are *legitimate* forms of representation, so that this more fundamental *representational issue* have not even appeared on the horizon (in spite of the fact that in this particular case Chomsky, from the very beginning, had emphasized the importance of generativity). Of course, one can excuse these *early* developments since at the time of their emergence, 1930s to 1950s, the serious representational issues, could not have appeared on the horizon; but now, at the beginning of the 21st century—when all kinds of applications, starting from general search engines (such as Google) and biological databases, all the way to various robotic applications "beg" for structural representations—the story is quite different.

Thus, the price of relying on conventional discrete representations is that the *more fundamental, underlying, representational formalism remained in the dark*, which, in turn, prevented the development of a scientifically much more satisfactory and definitive framework for computation, including the realization of the full potential of many interesting ideas such as, to take a recent example, the membrane computing.

For a discussion of the inherent, or structural limitations of the two main conventional formalisms—vector space, or numeric, and logical—see [4], [5]. Incidentally, it is these inherent limitation of the numeric (representational) formalism that, I believe, are responsible for the predominance of statistical over structural considerations in machine learning, pattern recognition, data mining, information retrieval, bioinformatics, cheminformatics, and many other applied areas. The question is not whether the *appropriate* statistical considerations should play some role in these areas, the obvious answer to which is "yes, of course", but, as I emphasized above, whether, *at present*—when we lack any satisfactory formalism for structural representation—we should be focusing on the development of new statistical techniques for the conventional formalisms that are *inherently* inadequate for dealing with classes (and hence, with the class-oriented needs of the above areas).

4   A Brief Sketch of the ETS Formalism

Unhappily or happily, the structure of this formalism has absolutely no analogues to compare it with, despite the fact that its main entities, "structs", may have a superficial resemblance to some other known discrete objects such as, for example, graphs, which is useful to keep in mind. Also, in view of limited space, in what follows I restrict myself to informal descriptions, while the formal definitions can be found in [6], Parts II and III. The most important point to keep in mind is that, in the formalism, all objects are *viewed and represented* as (temporal) structural processes, in which the basic units are *structured events*, each responsible for transforming the flow of several "regular" processes (see Figs. 1, 2).

4.1 Primitive Transformations

The *first basic concept* is that of a **primitive transformation,** or primitive *event,* or simply **primitive,** several of which are depicted in Fig. 1. In contrast to the basic units in conventional formalisms, this concept "atomic" to the ETS formalism is relatively non-trivial, *carrying both semantic and syntactic loads.* It stands for a fixed *"micro-event"* responsible for transforming one set of adjacent/interacting processes, called *initial processes,* into another set of processes, called *terminal processes* (in Fig. 2, both are shown as lines connecting primitives). In other words, the concept of primitive transformation encapsulates that of a "standard"/fixed interaction[7] of several processes, where each belongs to a particular class of primal processes, or a **primal class**[8]. One can assume that the processes involved are "periodic" and the event is responsible for their partial or complete modification into another set of periodic processes. The formal structure of the event is such that it does not depend on the concrete initial (or concrete terminal) process, as long as each of the processes involved belongs to the same (fixed) primal class of processes depicted in Fig. 1 as a small solid shape at the top (or at the bottom) of the event. As one can see, *at this, basic, (or 0th), stage of representation*[9], the structure of the initial and terminal processes is suppressed, as is the *internal* structure of the transforming event itself, and what's being captured by the formal structure is the "external" structure of the event.

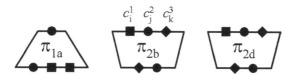

Fig. 1. Pictorial illustration of three primitives. The first subscript of a primitive refers to the class of primitives all sharing the same structure, e.g. π_{2b} and π_{2d}. The initial *classes* are marked as various shapes on the top, while the terminal *classes* are shown as shapes on the bottom: each shape stands for a particular class of processes. The only concrete processes—i.e. elements of these classes—identified in the figure as the initial processes of the primitive π_{2b} with label $b = \langle c_i^1, c_j^2, c_k^3 \rangle$, where c_t^s is the tth process from primal class C_s, $s = 1, 2, 3$.

Since all of nature is composed of various temporal processes, examples of the above events are all around us: e.g. elementary particle collision, formation of a two-cell blastula from a single cell (initial process is the original cell and the terminal processes are the resulting two cells), collision of two cars, the event associated with the transforming effect on the listener's memory of the sentence

---

[7] The *internal* structure of such event/interaction is undisclosed.

[8] The concept of class permeates all levels of consideration.

[9] In this paper, I discuss only a single-stage version of ETS. For *multi-stage* version see [6], Part IV.

"Alice and Bob had a baby" (initial processes are Alice and Bob and the terminal processes are Alice, Bob, and the baby), etc., where each mentioned event transforms the "flow" of the corresponding stable/regular processes involved.

4.2 Structs

The *second basic concept* is that of a **struct** formed by a (temporal) sequence of the above primitive events, as shown in Fig. 2. It should be easy to see how the temporal process of Peano construction of natural numbers (see Fig. 3) was generalized to the construction of structs: the single "structureless" unit out of which a number is built was replaced by multiple structural ones. An immediate and important consequence of the distinguishability (or multiplicity) of units in the construction process is that we can now see which unit was attached and when. Hence, the resulting (object) representation for the first time embodies both temporal and structural information in the form of a formative, or generative, object history recorded as a series of (structured) events. This was one of the basic driving motivations for the development of ETS, while the other main motivation was the development of a unified *class-oriented* representational formalism that would satisfy the needs of a large variety of information-processing areas as well as natural sciences.

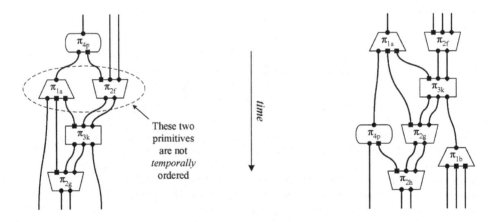

Fig. 2. Two illustrative examples of (short) structs

4.3 Struct Assembly

One of the basic *operations on structs* is **struct assembly** (as depicted in Fig. 4), which relies on the events shared by the structs involved. This operation allows one to combine in one struct several *separately observed*, but typically overlapping, structs (e.g. those representing several facial features such as the eyes and the nose). In particular, the overlap of several structs representing several actual processes captures a "non-interfering" *interaction* of these processes.

It is not difficult to see now the fundamental difference between a struct and, for example, a string: the main difference has to do with the temporal nature

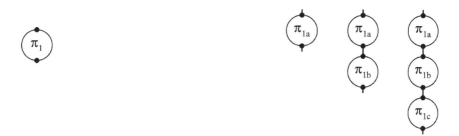

Fig. 3. The single primitive involved in the ETS representation of natural numbers, and three structs representing the numbers 1, 2, and 3

of ETS representation, which allows one to "compare" any objects *based on their "formative history"*. The latter information is simply not available in all conventional forms of representation.

Thus, to repeat, the ETS object representation captures the object's formative/generative history which, for the first time, brings considerable additional information into the object representation. I should add that, *from the applied point of view*, the adjective "formative" does not (and cannot) refer to the object's *actual* formative history, since, in many cases, it is not accessible to us, but rather to a particularly chosen application mode of approaching/recording it, i.e. to the mode of struct construction selected by the application developers.

4.4 Level 0 Classes

The *third basic concept* is that of a **class**. Even within the basic representational stage—the only one I discuss here—each class[10] is also viewed as possibly multi-leveled. A *single-level class representation* is specified by means of a **single-level, or level 0, class generating system**, which details the stepwise mode of construction of the class elements.[11] *Each (non-deterministic) step* in such a system is specified by a set of (level 0) *constraints*. Each constraint—also a major concept, which not introduced here—is a *formal* specification of a family of structs sharing some structural "components" in the form of common substructs. During a step in the construction process, the struct that is being attached (at this step) to the part of the class element that has been assembled so far must satisfy one of the constraints for this step. To be more accurate, in the definition, it is assumed that each such step can be preceded by a step executed by the "environment", i.e. by some other class generating system "intervening" or "participating" in the construction process (see Fig. 5). Thus, quite appropriately and realistically, a change in the environment may result in a class change,

---

[10] Here, I do not include the primal classes, which are assumed to be of undisclosed structure.

[11] Note that since we began with level 0, it cannot recurse to any lover level; level 1 is built on top of level 0; level 2 is built on top of level 1; etc. Also note the difference between a *level* and a *stage* (see the next subsection): levels refer to those of classes and appear within a single representational stage.

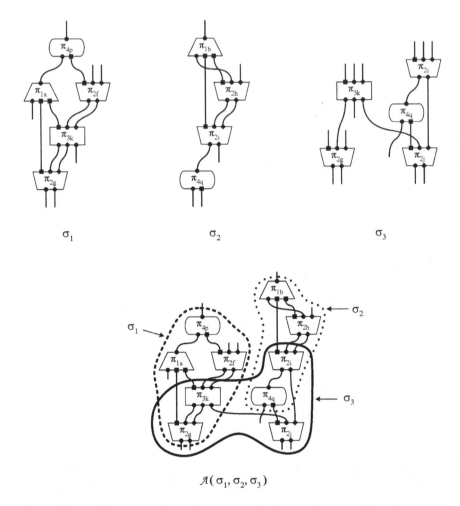

Fig. 4. Three structs and their assembly. Note that the second link connecting primitives π_{3k} and π_{2g} in the assembly comes from σ_1 (but not from σ_3, despite the delineation of σ_3 with a bold line).

without an attendant change in the class generating system itself. Such a concept of class admits the effects of the environment in a "natural" manner.

4.5   Level 1 Structs

Suppose that an agent has already learned several level 0 classes, which together form the current *level 0 class setting*. Then, when representing objects, the agent has an access to a more refined form of object representation than a plain level 0 struct: it can now see if this struct is in fact composed of several level 0 class elements (each belongs to one of the classes in level 0 class setting, see Fig. 6). This leads to the concept of the next level (level 1) struct, which provides extra

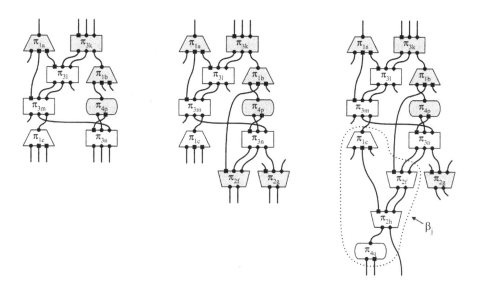

Fig. 5. Pictorial representation of a generic two-step generative "unit" in the construction of a class element: a step by the environment (bottom shaded primitives in the second struct, added to the first one) followed by a step made by the class generating system (substruct β_j in the third struct attached to the second struct)

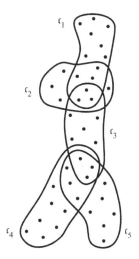

Fig. 6. Simplified pictorial representation of a level 1 struct in the contracted form: dots stand for primitives and solid lines delineate level 0 class elements c_i's

representational information as compared to the underlying level 0 struct in the form of the appropriate partition of the previous level (level 0) struct.

4.6   Higher Level Classes

In the (recursive) *k-level* version of the *class representation*, for $k \geq 2$, each step is associated with a set of *level* $(k-1)$ *constraints*. However, during the construction process, level $(k-1)$ struct that is being attached (at this step) to the previously constructed part must now be composed only out of level $(k-2)$ admissible class elements[12] in a manner satisfying one of the constraints for this step. Figure 7 illustrates this construction process for level 1 class element.

For a level 2 class element this element is an output of a level 2 (or three-levels) class generating system at each step of its construction, the relevant part is assembled out of several level 1 class elements in accordance with one of the constraints specified for this step.

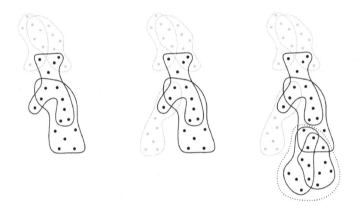

Fig. 7. Illustration of a generic two-step generative unit in the construction of some *level 1* class element. Dots stand for primitives, and solid lines delineate level 0 class elements, which now serve as simplest construction units. A step by the environment (bottom left gray addition in the second struct) is followed by a step made by a level 1class generating system itself (struct delineated by the dotted line).

5   Some Initial Computational Questions Arising Within the ETS Formalism

For simplicity, I discuss only a basic single-stage computational "agenda" within the ETS formalism, which can be stated in the following form: *Given an abstract delineation of a family of multi-level classes (or class generating systems) and given several 0-level ("training") structs from a class belonging to the given family, the goal is to "extract" the corresponding class generating system.*

In other words, I suggest that within such (inductive) framework, the computational theory is aimed at developing the theory and techniques that would allow one, first, to preprocess (or partition, or "mark up") the input structs in

---

[12] Each of those must come from a class belonging to a (previously learned or given) set of level $(k-2)$ classes, comprising *level* $(k-2)$ *class setting*.

a consistent manner, based on the structure of the given family of multi-level classes; and then one can proceed to the extraction of the underlying class generating system.

So first of all, one would need to develop a corresponding structural theory, which could be thought of as a "structural" generalization of the Chomsky's language hierarchy. As in the case of Chomsky hierarchy, such a theory would allow one to restrict the above problem to a particular (structural) family of classes. The development of such a hierarchy should probably proceed along the lines somewhat similar to those followed by Chomsky, with the following quite natural adjustment: within the hierarchy, each family of classes should be introduced via the appropriate restriction on the set of admissible constraints—which are the ETS analogues of the production rules—involved in the specification of the class. However, it is also important to emphasize the differences between the ETS and the conventional string (and graph hierarchies), and such differences concern not just the presence of levels in ETS. When defining a conventional, for example Chomsky, hierarchy, *one is not treating a string as a form of object representation* as understood above. This results, on the one hand, in a wider range of admissible production rules (not all are meaningful from the representational point of view), and on the other hand, in a variety of qualitatively different machines associated with each family of languages (e.g. deterministic/nondeterministic). There are reasons to believe that, in the case of ETS formalism, in view of the temporal form of representation, the situation is more palatable.

Next, based on such a structure theory, one would need to develop techniques for the above *class related* (structural) preprocessing, or partitioning, of the structs in a given (training) set of structs. Even this task is highly non-trivial, since the tiling of each given struct must be accomplished relying on the *admissible* previous level class elements only (see Fig. 6). The final stage is related to the construction of the corresponding class representation.

It is understood that, in the meantime, to proceed with various applications, one does not need to wait for the development of a full-fledged structural theory, since in each concrete application, the corresponding family of classes can readily be specified.X

Thus, it should be clear that in contrast to the conventional computational models—where, as was mentioned above, the defining agenda was logical—I propose that it is the (structural) inductive agenda that should now drive the development of the computational framework. This proposal is not really surprising, given that the new framework is aimed at supporting a continuous dynamic interaction between the "machine" and various *natural environments*, where, as was hypothesized above, the temporal/structural nature of object representation is ubiquitous.

6   Conclusion

I hinted that it might be prudent for some researchers in computation to add to their research interests a new area that we called *Inductive Informatics*, which can be thought of as a reincarnation of a four-hundred-year-old Francis Bacon's project of putting all sciences on firm inductive footing, with the attendant restructuring of the basic formal language and methodology. In hindsight, it appears that the major obstacle has been the lack of a classification-oriented—or which appears to be the same thing "structural"—representational formalism. Our proposed version of such formalism is a far-reaching, structural, generalization of the numeric representation, and in that sense *it is much closer to mathematics than the logic*, which was suggested by von Neumann [2] to be a desirable direction. Moreover, with the appearance of such a formalism, the concept of *class representation*, for the first time, becomes meaningful, and it also becomes quite clear that it is the *intrinsic* incapability of the conventional representational formalisms to support this fundamental concept that is responsible for our previous failure to realize Bacon's vision.

In addition to the new theoretical horizons that are being opened up within new (event-based) representational formalisms, the other main reason why such formalisms should be of interest to researchers in computation has to do with the *immediate* practical benefits one can derive from their various applications in machine learning, pattern recognition, data mining, information retrieval, bioinformatics, cheminformatics, and many other applied areas. The validity of the last statement should become apparent if one followed carefully the above point regarding the new and rich aspect of object representation—i.e. the object's formative history—that now becomes available within the ETS formalism.

Acknowledgment. I thank Oleg Golubitsky for a discussion of the paper and Ian Scrimger for help with formatting.

References

1. Cook, S.A.: The complexity of theorem proving. In: Laplante, P. (ed.) Great Papers in Computer Science, vol. 2, West Publishing Co. St. Paul, Minneapolis (1996)
2. Neumann, von J.: The general and logical theory of automata. In: Pylyshyn, Z.W. (ed.) Perspectives in the Computer Revolution, pp. 99–100. Prentice-Hall, Englewood Cliffs, New Jersey (1970)
3. Turing, A.M.: On computable numbers, with an application to the Entscheidungsproblem. In: Laplante, P. (ed.) Great Papers in Computer Science, pp. 303–304. West Publishing Co. St. Paul, Minneapolis (1996)
4. Goldfarb, L.: Representational formalisms: What they are and why we haven't had any. In: Goldfarb, L. (ed.) What Is a Structural Representation (in preparation) http://www.cs.unb.ca/~goldfarb/ETSbook/ReprFormalisms.pdf
5. Goldfarb, L.: On the Concept of Class and Its Role in the Future of Machine Learning. In: Goldfarb, L. (ed.) What Is a Structural Representation (in preparation) http://www.cs.unb.ca/~goldfarb/ETSbook/Class.pdf
6. Goldfarb, L., Gay, D., Golubitsky, O., Korkin, D.: What is a structural representation? A proposal for an event-based representational formalism. In: Goldfarb, L. (ed.) What Is a Structural Representation (in preparation) http://www.cs.unb.ca/~goldfarb/ETSbook/ETS5.pdf

Computing Through Gene Assembly

Tseren-Onolt Ishdorj[1,3] and Ion Petre[1,2]

[1] Computational Biomodelling Laboratory
Department of Information Technologies
Åbo Akademi University, Turku 20520 Finland
{tishdorj,ipetre}@abo.fi
[2] TUCS, Turku 20520 Finland
[3] Research Group on Natural Computing
Department of CS and AI, Sevilla University
Avda. Reina Mercedes s/n, 41012 Sevilla, Spain

Abstract. The intramolecular gene assembly model, [1], uses three molecular recombination operations ld, dlad, and hi. A computing model with two contextual recombination operations del and trl, which are based on ld and dlad, respectively, is considered in [6] and its computational power is investigated. In the present paper, we expand the computing model with a new molecular operation such as cpy - *copy*. Then we prove that the extended contextual intramolecular gene assembly model is both computationally universal and efficient.

1 Introduction

There have been proposed two formal models to explain the gene assembly process in ciliates: intermolecular model, for instance in [8], and intramolecular model, for instance in [1]. They both are based on so called "pointers" - short nucleotide sequences (about 20 bp) lying on the borders between coding and non-coding blocks. Each next coding block starts with a pointer-sequence repeating exactly the pointer-sequence in the end of the preceding codding block from the assembled gene. It is supposed that the pointers guide the alignment of coding blocks during the gene assembly process. The intramolecular model proposes three operations: ld (loop with direct pointers), hi (hairpin loop with inverted pointers), dlad (double loop with alternating direct pointers).

The context sensitive variants of the intramolecular operations dlad and ld have been considered in [6]. The accepting contextual intramolecular recombination (AIR) system using the operations *translocation* based on (dlad) and *deletion* based on (ld) has been proved to be equivalent in power with Turing machine.

In the present paper, we expand the set of contextual intramolecular operations with a new operation: *copy* cpy_p. Intramolecular *copy* operation is first considered in [14]. The extended intramolecular recombination model with three types of contextual operations *translocation, deletion,* and *copy* computing along a single string is as powerful as Turing machine, and efficient enough to solve **NP**-complete problems (in particular, SAT) in polynomial time.

S.G. Akl et al.(Eds.): UC 2007, LNCS 4618, pp. 91–105, 2007.
© Springer-Verlag Berlin Heidelberg 2007

No ciliate-based efficient algorithm for intractable problems has been presented in the literature so far. This makes the efficiency result in this paper novel in the theory of ciliate computation.

A particularly interesting feature of the system is that the recombination operations are applied in a maximally parallel manner.

To solve an instance of SAT, we encode the problem into a string. Then during the recombination steps, the encoded problem (propositional formula) is duplicated by *copy* operation along the string. Meanwhile, the truth-assignments attached to the propositional formula are constructed. In the end, all possible truth-assignments are generated in a linear number of steps, and each truth-assignment is checked to see whether it satisfies the encoded formula.

The universality result of [6] uses a number of copies of a given substring along the working string in order to start a computation. By extension of the model with the *copy* operation, the necessary substrings will be generated during the computation instead of making the working string as full storage in advance as in [6]. This saves on the size and complexity of the encoding.

2 Preliminaries

We assume the reader to be familiar with the basic elements of formal languages and Turing computability [13], DNA computing [12], and computational complexity [10]. We present here only some of the necessary notions and notation.

Using an approach developed in a series of works (see [9], [11], [2], and [7]) we use *contexts* to restrict the application of molecular recombination operations, [12], [1]. First, we give the formal definition of splicing rules. Consider an alphabet Σ and two special symbols, $\#, \$ \notin \Sigma$. A *splicing rule* (over Σ) is a string of the form $r = u_1 \# u_2 \$ u_3 \# u_4$, where $u_1, u_2, u_3, u_4 \in \Sigma^*$. (For a maximal generality, we place no restriction on the strings u_1, u_2, u_3, u_4. The cases when $u_1 u_2 = \lambda$ or $u_3 u_4 = \lambda$ could be ruled out as unrealistic.)

For a splicing rule $r = u_1 \# u_2 \$ u_3 \# u_4$ and strings $x, y, z \in \Sigma^*$ we write $(x, y) \vdash_r z$ if and only if $x = x_1 u_1 u_2 x_2$, $y = y_1 u_3 u_4 y_2$, $z = x_1 u_1 u_4 y_2$, for some $x_1, x_2, y_1, y_2 \in \Sigma^*$. We say that we *splice* x, y at the *sites* $u_1 u_2$, $u_3 u_4$, respectively, and the result is z. This is the basic operation of DNA molecule recombination.

A splicing scheme [5] is a pair $R = (\Sigma, \sim)$, where Σ is the alphabet and $\sim$, the pairing relation of the scheme, $\sim \subseteq (\Sigma^+)^3 \times (\Sigma^+)^3$. Assume we have two strings x, y and a binary relation between two triples of nonempty words $(\alpha, p, \beta) \sim (\alpha', p, \beta')$, such that $x = x' \alpha p \beta x''$ and $y = y' \alpha' p \beta' y''$; then, the strings obtained by the recombination in the context from above are $z_1 = x' \alpha p \beta' y''$ and $z_2 = y' \alpha' p \beta x''$. When having a pair $(\alpha, p, \beta) \sim (\alpha', p, \beta')$ and two strings x and y as above, $x = x' \alpha p \beta x''$ and $y = y' \alpha' p \beta' y''$, we consider just the string $z_1 = x' \alpha p \beta' y''$ as the result of the recombination (we call it one-output-recombination), because the string $z_2 = y' \alpha' p \beta x''$, we consider as the result of the one-output-recombination with the respect to the symmetric pair $(\alpha', p, \beta') \sim (\alpha, p, \beta)$.

A rewriting system $M = (S, \Sigma \cup \{\#\}, P)$ is called a *Turing machine* (we use also abbreviation TM), [13], where: (*i*) S and $\Sigma \cup \{\#\}$, where $\# \notin \Sigma$ and $\Sigma \neq \emptyset$, are two disjoint sets referred to as the *state* and the *tape* alphabets; we fix a symbol from Σ, denote it as $\sqcup$ and call it "blank symbol"; (*ii*) Elements s_0 and s_f of S are the *initial* and the *final* states respectively; (*iii*) The productions (rewriting rules) of P are of the forms

(1) $s_i a \longrightarrow s_j b$; (2) $s_i a c \longrightarrow a s_j c$; (3) $s_i a \# \longrightarrow a s_j \sqcup \#$; (4) $c s_i a \longrightarrow s_j c a$; (5) $\# s_i a \longrightarrow \# s_j \sqcup a$; (6) $s_f a \longrightarrow s_f$; (7) $a s_f \longrightarrow s_f$, where s_i and s_j are states in S, $s_i \neq s_f$, and a, b, c are in Σ.

The TM M changes from one configuration to another one according to its set of rules P. We say that the Turing machine M *halts* with a word w if there exists a computation such that, when started with the read/write head positioned at the beginning of w, the TM eventually reaches the final state, i.e., if $\# s_0 w \#$ derives $\# s_f \#$ by successive applications of the rewriting rules (1)–(7) from P. The language $L(M)$ *accepted* by the TM M is the set of words on which M halts. If TM is deterministic, then there is the only computation possible for each word. The family of languages accepted by Turing machines is equivalent to the family of languages accepted by deterministic Turing machines.

The satisfiability problem, SAT, is a well known **NP**-complete problem. It asks whether or not for a given propositional formula in the conjunctive normal form there is a truth-assignment of variables such that the formula assumes the value *true*. The details of SAT are considered in Section 4.

3   The Contextual Intramolecular Operations

The contextual intramolecular *translocation* and *deletion* operations are the generalizations of dlad and ld operations, respectively, as defined in [6]. We consider a splicing scheme $R = (\Sigma, \sim)$. Denote $core(R) = \{p \mid (\alpha, p, \beta) \sim (\gamma, p, \delta) \in R$ for some $\alpha, \beta, \gamma, \delta \in \Sigma^+\}$.

Definition 1. *The contextual intramolecular translocation operation with respect to R is defined as* $\mathrm{trl}_{p,q}(xpuqypvqz) = xpvqypuqz$, *where there are such relations* $(\alpha, p, \beta) \sim (\alpha', p, \beta')$ *and* $(\gamma, q, \delta) \sim (\gamma', q, \delta')$ *in R, that* $x = x'\alpha$, $uqy = \beta u' = u''\alpha'$, $vqz = \beta' v'$, $xpu = x''\gamma$, $ypv = \delta y' = y''\gamma'$ *and* $z = \delta' z'$.

We say that operation $\mathrm{trl}_{p,q}$ is applicable, if the contexts of the two occurrences of p as well as the contexts of the two occurrences of q are in the relation $\sim$. Substrings p and q we call *pointers*. In the result of application of $\mathrm{trl}_{p,q}$ strings u and v, each flanked by pointers p and q, are swapped. If from the non-empty word u we get by $\mathrm{trl}_{p,q}$ operation word v, we write $u \Longrightarrow_{\mathrm{trl}_{p,q}} v$ and say that v is obtained from u by $\mathrm{trl}_{p,q}$ operation. The *translocation* is depicted in Fig. 1.

Definition 2. *The contextual intramolecular deletion operation with respect to R is defined as* $\mathrm{del}_p(xpupy) = xpy$, *where there is a relation* $(\alpha, p, \beta) \sim (\alpha', p, \beta')$ *in R that* $x = x'\alpha$, $u = \beta u' = u''\alpha'$, *and* $y = \beta'y'$.

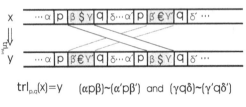

$\mathsf{trl}_{p,q}(x) = y$ $(\alpha p\beta) \sim (\alpha'p\beta')$ and $(\gamma q\delta) \sim (\gamma'q\delta')$

Fig. 1. Translocation operation

In the result of applying del_p, the string u flanked by two occurrences of p is removed, provided that the contexts of those occurrences of p are in the relation $\sim$. If from a non-empty word u we obtain a word v by del_p, we write $u \Longrightarrow_{\mathsf{del}_p} v$ and say that the word v is obtained from u by del_p operation. See Fig. 2 A.

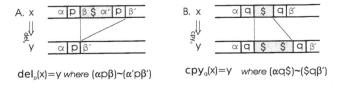

$\mathsf{del}_p(x) = y$ where $(\alpha p\beta) \sim (\alpha'p\beta')$ $\mathsf{cpy}_q(x) = y$ where $(\alpha q\$) \sim (\$q\beta')$

Fig. 2. A. Deletion and B. Copy operations

We introduce here an additional operation *the contextual intramolecular copy*.

Definition 3. *The contextual intramolecular copy operation with respect to R is defined as* $\mathsf{cpy}_q(xquqy) = xquuqy$, *where there is a relation* $(\alpha, q, \beta) \sim (\alpha', q, \beta')$ *in R that* $x = x'\alpha$, $u = \beta = \alpha'$, $y = \beta'y'$.

As a result of applying cpy_q, the substring u flanked by two occurrences of q is duplicated. The substring u supposed to be duplicated itself must be also the following substring of the first pointer q and the preceding substring of the second occurrence of q. If from the non-empty word u we obtain a word v by cpy_q, we write $u \Longrightarrow_{\mathsf{cpy}_q} v$ and say that the word v is obtained from u by cpy_q operation. The *copy* operation is depicted in Fig. 2 B.

We use the next notations:

$$\mathsf{trl} = \{\mathsf{trl}_{p,q} \mid p, q \in core(R)\},$$
$$\mathsf{del} = \{\mathsf{del}_p \mid p \in core(R)\},$$
$$\mathsf{cpy} = \{\mathsf{cpy}_p \mid p \in core(R)\}.$$

Then we define the set of all contextual intramolecular operations as follows: $\widetilde{R} = \mathsf{trl} \cup \mathsf{del} \cup \mathsf{cpy}$.

Definition 4. *Let a triplet* $R_{op} = (\Sigma, \sim, op)$ *be a splicing scheme that an operation* $op \in \widetilde{R}$ *performs only in the contexts defined by the set of splicing relations in R_{op}. We say that R_{op} is an operation class of the splicing scheme $\sim$ for the operation* $op \in \widetilde{R}$.

We consider the parallelism for the intramolecular recombination model. Intuitively, a number of operations can be applied in parallel to a string if the applicability of each operation is independent of the applicability of the other operations. The parallelism in intramolecular gene assembly was initially studied in a different setting in [4]. We recall the definition of parallelism following [4] with a small modification adapted to our model.

Definition 5. *Let $S \subseteq R$ be a set of k rules and let u be a string. We say that the rules in S can be applied in parallel to u if for any ordering $\varphi_1, \varphi_2, \ldots, \varphi_k$ of S, the composition $\varphi_k \circ \varphi_{k-1} \circ \cdots \circ \varphi_1$, is applicable to u.*

In our proof below we use a different notion of parallelism. We introduce it here is the form of the *maximally parallel* application of a rule to a string. First, we define the working places of a operation $\varphi \in \widetilde{R}$ on a given string where φ is applicable.

Definition 6. *Let w be a string. The working places of a operation $\varphi \in \widetilde{R}$ for w is a set of substrings of w written as $Wp(\varphi(w))$ and defined by*

$$Wp(\mathsf{trl}_{p,q}(w)) = \{(w_1, w_2) \in Sub(w) \mid \mathsf{trl}_{p,q}(xw_1yw_2z) = xw_2yw_1z\}.$$
$$Wp(\mathsf{del}_p(w)) = \{w_1 \in Sub(w) \mid \mathsf{del}_p(xpw_1py) = xpy\}.$$
$$Wp(\mathsf{cpy}_p(w)) = \{w_1 \in Sub(w) \mid \mathsf{cpy}_p(xw_1y) = xw_1w_1y\}.$$

Definition 7. *Let w be a string. The smallest working places of an operation $\varphi \in \widetilde{R}$ for w is a subset of $Wp(\varphi)(w)$ written as $Wp_s(\varphi(w))$ and defined by*

$$\begin{aligned}
Wp_s(\mathsf{trl}_{p,q}(w)) = \{&(w_1, w_2) \in Wp(\mathsf{trl}_{p,q}(w)) \mid \text{ for all } w_1' \in Sub(w_1) \text{ and}\\
&w_2' \in Sub(w_2) \text{ and } (w_1', w_2') \neq (w_1, w_2),\\
&(w_1', w_2') \notin Wp(\mathsf{trl}_{p,q}(w))\}.\\
Wp_s(\mathsf{del}_p(w)) = \{&w_1 \in Wp(\mathsf{del}_p(w)) \mid \text{ for all } w_1' \in Sub(w_1),\\
&\text{and } w_1' \neq w_1, w_1' \notin Wp(\mathsf{del}_p(w))\}.\\
Wp_s(\mathsf{cpy}_p(w)) = \{&w_1 \in Wp(\mathsf{cpy}_p(w)) \mid \text{ for all } w_1' \in Sub(w_1),\\
&\text{and } w_1' \neq w_1, w_1' \notin Wp(\mathsf{cpy}_p(w))\}.
\end{aligned}$$

Definition 8. *Let Σ be a finite alphabet and $\widetilde{R}$ the set of rules defined above. Let $\varphi \in R$ and $u \in \Sigma^*$. We say that $v \in \Sigma^*$ is obtained from u by applying φ in a maximally parallel way, denoted $u \Longrightarrow_{\varphi}^{max} v$, if $u = \alpha_1 u_1 \alpha_2 u_2 \ldots \alpha_k u_k \alpha_{k+1}$, and $v = \alpha_1 v_1 \alpha_2 v_2 \ldots \alpha_k v_k \alpha_{k+1}$, where $u_i \in Wp_s(\varphi)(w)$ for all $1 \leq i \leq k$, and also, $\alpha_i \notin Wp(\varphi(w))$, for all $1 \leq i \leq k+1$.*

Note that a rule $\varphi \in \widetilde{R}$ may be applied in parallel to a string in several different ways, as shown in the next example.

Example 1. Let $\mathsf{trl}_{p,q}$ be the contextual translocation operation applied in the context $(x_1, p, x_2) \sim (x_3, p, x_4)$ and $(y_1, q, y_2) \sim (y_3, q, y_4)$. We consider the string $u = x_1 p x_2 \$_1 y_1 q y_2 \$_2 x_3 p x_4 \$_3 x_3 p x_4 \$_4 y_3 q y_4$.

Note that there are two occurrences of p with context x_3 and x_4. We can obtain two different strings from u by applying $\mathsf{trl}_{p,q}$ in a parallel way as follows:

$$u \Longrightarrow_{\mathsf{trl}_{p,q}}^{max} v' \text{ where } v' = x_1px_4\$_3x_3px_4\$_4y_3qy_2\$_2x_3px_2\$_1y_1qy_4.$$

$$u \Longrightarrow_{\mathsf{trl}_{p,q}}^{max} v'' \text{ where } v'' = x_1px_4\$_4y_3qy_2\$_2x_3px_4\$_3x_3px_2\$_1y_1qy_4.$$

Here only the second case satisfies the definition of maximally parallel application of $\mathsf{trl}_{p,q}$ because it applies for the smallest working place.

Example 2. Let del_p be the contextual deletion operation applied in the relation of $(x_1x_2, p, x_3) \sim (x_3, p, x_1)$, and consider the string $u = x_1x_2px_3px_1x_2p\ x_3px_1$. The unique correct result obtained by maximally parallel application of del_p to u is $x_1x_2px_3px_1x_2px_3px_1 \Longrightarrow_{\mathsf{del}_p}^{max} x_1x_2px_1x_2px_1.$

4   Computational Efficiency and Universality

Definition 9. *An extended accepting intramolecular recombination (eAIR) system is a tuple $G = (\Sigma, \sim, \widetilde{R}, \alpha_0, w_t)$ where $\widetilde{R}$ is the set of recombination operations, $\alpha_0 \in \Sigma^*$ is the start word, and $w_t \in \Sigma^+$ is the target word. If α_0 is not specified, we omit it. The operation classes of splicing scheme $R_{op} = (\Sigma, \sim, op), op \in \widetilde{R}$ are defined. Then the operations are applied in a sequentially or a maximally parallel manner according to the operation classes. The language accepted by G is defined by $L(G) = \{w \in \Sigma^* \mid \alpha_0w \Longrightarrow_{R_{op \in \widetilde{R}}}^* w_t\}.$*

We use the extended accepting intramolecular recombination (eAIR) systems as decision problem solvers. A possible correspondence between decision problems and languages can be done via an encoding function which transforms an instance of a given decision problem into a word, see, e.g., [3].

Definition 10. *We say that a decision problem X is solved in time $O(t(n))$ by extended accepting intramolecular recombination systems if there exists a family $\mathcal{A}$ of extended AIR systems such that the following conditions are satisfied:*

1. *The encoding function of any instance x of X having size n can be computed by a deterministic Turing machine in time $O(t(n))$.*
2. *For each instance x of size n of the problem one can effectively construct, in time $O(t(n))$, an extended accepting intramolecular recombination system $G(x) \in \mathcal{A}$ which decides, again in time $O(t(n))$, the word encoding the given instance. This means that the word is accepted if and only if the solution to the given instance of the problem is* YES.

Theorem 1. SAT *can be solved deterministically in linear time by an extended accepting intramolecular recombination system constructed in polynomial time in the size of the given instance of the problem.*

Proof. Let us consider a propositional formula in the conjunctive normal form, $\varphi = C_1 \wedge \cdots \wedge C_m$, such that each clause $C_i, 1 \leq i \leq m$, is of the form $C_i = y_{i,1} \vee \cdots \vee y_{i,k_i}, k_i \geq 1$, where $y_{i,j} \in \{x_k, \bar{x}_k \mid 1 \leq k \leq n\}$.

We construct an extended accepting intramolecular recombination system

$$G = (\Sigma, \sim, \widetilde{R}, \text{YES}), \quad \text{where}$$

$$\Sigma = \{\$_i \mid 0 \leq i \leq m+1\} \cup \{x_i, x_{\bar{i}}, \langle_i, \langle_{\bar{i}}, \rangle_i, \rangle_{\bar{i}}, \dagger_i, \dagger_{\bar{i}} \mid 1 \leq i \leq n\}$$
$$\cup \{f_i \mid 0 \leq i \leq n+1\} \cup \{T, F, \vee, \text{Y}, \text{E}, \text{S}\},$$
$$R = (\Sigma, \sim),$$
$$\widetilde{R} = \{\text{trl}_{\langle_i, \rangle_i}, \text{trl}_{\langle_{\bar{i}}, \rangle_{\bar{i}}}, \text{del}_{f_i}, \text{del}_{\dagger_i}, \text{del}_{\dagger_{\bar{i}}} \mid \langle_i, \rangle_i, \langle_{\bar{i}}, \rangle_{\bar{i}}, f_i, \dagger_i, \dagger_{\bar{i}} \in core(R), 1 \leq i \leq n\}$$
$$\cup \{\text{cpy}_{f_i} \mid f_i \in core(R), 0 \leq i \leq n-1\} \cup \{\text{del}_{\$_i} \mid \$_i \in core(R), 1 \leq i \leq m\}$$
$$\cup \{\text{del}_{\text{E}} \mid \text{E} \in core(R)\},$$

and the operation classes R_{trl}, R_{del}, and R_{cpy} are defined below.

We encode each clause C_i as a string bounded by $\$_i$ in the following form: $c_i = \$_i \vee \langle_{\sigma(j_b)} x_{\sigma(j_b)} \rangle_{\sigma(j_b)} \vee \cdots \vee \langle_{n_{\sigma(j_b)}} x_{n_{\sigma(j_b)}} \rangle_{n_{\sigma(j_b)}} \vee \$_i$, where $b \in \{0,1\}, \sigma(j_1) = j, \sigma(j_0) = \bar{j}$, and x_j stands for variable x_j, while $x_{\bar{j}}$ stands for negated variable $\bar{x}_j, 1 \leq j \leq n$, in the formula φ. The instance φ is encoded as:

$$\delta = \$_0 c_1 \ldots c_m \$_{m+1}.$$

In order to generate all possible truth-assignments for all variables $x_1, x_2, \ldots, x_n$ of the formula, we consider a string of the form

$$\gamma = \dagger_1 \langle_1 T \rangle_1 \dagger_1 \dagger_{\bar{1}} \langle_{\bar{1}} F \rangle_{\bar{1}} \dagger_{\bar{1}} \cdots \dagger_n \langle_n T \rangle_n \dagger_n \dagger_{\bar{n}} \langle_{\bar{n}} F \rangle_{\bar{n}} \dagger_{\bar{n}}.$$

Here T and F denote the truth-values *true* and *false*, respectively. Then m copies of γ are attached to the end of the encoded formula, leading to $\beta = \delta \gamma^m$.

If we have a propositional formula $\varphi' = (x_1 \vee \bar{x}_2) \wedge (x_1 \vee x_2)$, then

$$\delta' = \$_0 \$_1 \vee \langle_1 x_1 \rangle_1 \vee \langle_{\bar{2}} x_{\bar{2}} \rangle_{\bar{2}} \vee \$_1 \$_2 \vee \langle_1 x_1 \rangle_1 \vee \langle_2 x_2 \rangle_2 \vee \$_2 \$_3, \quad \text{and}$$
$$\gamma' = \dagger_1 \langle_1 T \rangle_1 \dagger_1 \dagger_{\bar{1}} \langle_{\bar{1}} F \rangle_{\bar{1}} \dagger_{\bar{1}} \dagger_2 \langle_2 T \rangle_2 \dagger_2 \dagger_{\bar{2}} \langle_{\bar{2}} F \rangle_{\bar{2}} \dagger_{\bar{2}}.$$

We use the following notations:

$$\gamma = \gamma_1 \gamma_2 \ldots \gamma_n \text{ where } \gamma_i = \gamma_i^{(1)} \gamma_i^{(0)}, \gamma_i^{(1)} = \dagger_i \langle_i T \rangle_i \dagger_i, \gamma_i^{(0)} = \dagger_{\bar{i}} \langle_{\bar{i}} F \rangle_{\bar{i}} \dagger_{\bar{i}}.$$
$$\gamma_{i,n} = \gamma_i \gamma_{i+1} \ldots \gamma_n . \gamma_{i,j}^{(b_{i,j})} = \gamma_i^{(b_i)} \gamma_{i+1}^{(b_{i+1})} \ldots \gamma_j^{(b_j)}, b_{i,j} = b_i b_{i+1} \ldots b_j \in \{0,1\}^+,$$
$$1 \leq i < j \leq n, \gamma_{1,1} = \gamma_1, \gamma_{n,n} = \gamma_n, \gamma_{1,0} = \lambda, \gamma_{n+1,n} = \lambda, \Gamma \in Sub(\gamma),$$
$$\alpha_{i,n} = f_i f_{i+1} \ldots f_n, \bar{\alpha}_{i,n} = f_n f_{n-1} \ldots f_i, 0 \leq i \leq n,$$
$$\alpha_{n+1,n+1} = \bar{\alpha}_{n+1,n+1} = f_{n+1}, \sigma(i_1) = i, \ \sigma(i_0) = \bar{i}, B \in \{T, F\}.$$

The input string π_1 (we call it in-string) containing the encoded propositional formula φ is of the form:

$$\pi_1 = \text{YE}\bar{\alpha}_{1,n+1} f_0 \alpha_{1,n+1} \beta \bar{\alpha}_{1,n+1} f_0 \alpha_{1,n+1} \text{ES}.$$

The size of the input is quadratic, $|\pi_1| \leq 4nm + 14n + 3m + 12$. Hence the system is constructible by a deterministic Turing machine in time $O(nm)$ in the size of the given instance of the problem.

Roughly speaking, the main idea of the algorithm by which the eAIR system solves SAT is as follows: (i) We generate all possible truth-assignments on the in-string according to the variables of the given instance of the propositional formula, while the encoded propositional formula with the attached Γ is copied, in linear time. (ii) Then each truth-assignment is assigned to its attached formula in the maximally parallel way. Step (iii), we check the satisfiability of the formula with regard to the truth-assignments. (iv) Finally, eAIR system decides to accept or not the input string π_1. We stress here that the recombination steps are guided very much by the classes of splicing schemes and by the contexts of rules defined in the splicing relations. The phases (i), (ii), (iii), and (iv) of the computation are illustrated in Fig. 3.

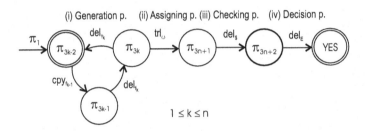

Fig. 3. A scheme of the algorithm

Let us start the computation accomplishing the above steps.

(i) Generating truth-value assignments. The generation of all possible truth-assignments takes $3n$ steps. The computation starts applying operation cpy_{f_0} (it is easy to observe that no other operations are applicable on the in-string at the first step), and then $\text{del}_{\dagger_1}$ and $\text{del}_{\dagger_{\bar{1}}}$ are applied parallel, following this step operation del_{f_1} applies. The next combination of three modules are applied automatically one after other in a cyclic order:

$$\text{cpy}_{f_i} \implies \text{del}_{\dagger_{i+1}}, \text{del}_{\dagger_{\overline{i+1}}} \implies \text{del}_{f_{i+1}}, 0 \leq i \leq n-1.$$

We emphasize here that the repeated pointers used in cpy_{f_i} are the closest to each other, because of the maximal parallel application of the rule and the definition of the operation cpy_p.

(1) The copy operation cpy_{f_i} duplicates the substrings s_i flanked by the repeats of f_i.

$$\alpha f_i s_i f_i \alpha' \implies_{\text{cpy}_{f_i}}^{max} \alpha f_i s_i s_i f_i \alpha', \ \alpha, \alpha' \in \Sigma^*, 0 \leq i \leq n-1.$$

The contexts for the cpy_{f_i} operation to be applied are

$-\ (\bar{\alpha}_{i+1,n+1}, f_i, s_i) \sim (s_i, f_i, \alpha_{i+1,n+1}) \in R_{\mathsf{cpy}}\,,$

where $s_i = \alpha_{i+1,n+1}\delta(\gamma_{1,i}^{(b_1,i)}\gamma_{i+1,n})^m\bar{\alpha}_{i+1,n+1}.$

At each $3k-2$th $(1 \le k \le n)$ step, the in-string contains substrings of the form:

$$f_{n+1}f_n \cdots f_{i+1}\underline{f_i}f_{i+1} \cdots f_n f_{n+1}\delta(\gamma_{1,i}^{(b_1,i)}\gamma_{i+1,n})^m f_{n+1}f_n \cdots f_{i+1}\underline{f_i}f_{i+1} \cdots f_n f_{n+1}.$$

Each substring s_i flanked by repeated f_i contains only one $\delta\Gamma^m$ part. The only possible rule to be applied in the context of this substring is cpy_{f_i} for the repeats $f_i, i = 3k-3$ (at this step there no rule is applicable from each R_{trl} and R_{del}, which can be checked out later on). No copy using $\mathsf{cpy}_{f_j}, j \ne 3k-3, j \ge 0$, may occur, because no $f_j, j < 3k-3$, has remained in the string; on the other hand, at this moment there is no pattern of the form $f_{j+1}f_j f_{j+1}, j > 3k-3$, to which cpy_{f_j} is applicable.

The copy cpy_{f_i} is applied in a maximally parallel way on the string for its smallest working places. As the result of this step, each substring flanked by f_i is copied as the following form:

$$f_{i+1}\underline{f_i}f_{i+1} \cdots f_{n+1}\delta(\gamma_{1,i}^{(b_1,i)}\gamma_{i+1,n})^m f_{n+1} \cdots f_{i+1}$$
$$f_{i+1} \cdots f_{n+1}\delta(\gamma_{1,i}^{(b_1,i)}\gamma_{i+1,n})^m f_{n+1} \cdots f_{i+1}\underline{f_i}f_{i+1}.$$

The substring flanked by f_i-s satisfies as the smallest working place for cpy_{f_i}, but cpy_{f_i} still can not apply to it because, on the one hand, by the definition of cpy_{f_i}, it applies to a substring s_i flanked by f_i-s and on the other hand, the substring s_i has to contain only one $\delta\Gamma^m$ part as the relations in R_{cpy} constrain it. But it is not the case with the above smallest working place. There are two $\delta\Gamma^m$ parts between the repeats of f_i. There are splicing relations for the repeats of f_i in R_{del} but we will check later that which are not satisfied on the current string.

(2) Thus, the next deletion operations $\mathsf{del}_{\dagger_i}$ and $\mathsf{del}_{\dagger_{\bar{i}}}$ apply to the current in-string. It is the $3k-1$th $(1 \le k \le n)$ step. The deletion $\mathsf{del}_{\dagger}$ is applied maximally parallel on the 2^k copies of Γ^m. By deletion $\mathsf{del}_{\dagger_i}$ (resp. $\mathsf{del}_{\dagger_{\bar{i}}}$), the truth-values $\langle_i T\rangle_i$ (resp. $\langle_{\bar{i}}F\rangle_{\bar{i}}$) are deleted wherever the contexts of $\mathsf{del}_{\dagger}$ are satisfied.

$$\alpha\dagger_{\sigma(i_b)}\langle_{\sigma(i_b)}B\rangle_{\sigma(i_b)}\dagger_{\sigma(i_b)}\alpha' \Longrightarrow_{\mathsf{del}_{\dagger_{\sigma(i_b)}}}^{max} \alpha\dagger_{\sigma(i_b)}\alpha', \alpha, \alpha' \in \Sigma^*, 1 \le i \le n.$$

$\mathsf{del}_{\dagger_{\bar{i}}} :\ (f_{i-1}\alpha_{i,n+1}\delta(\gamma_{1,i-1}^{(b_1,i-1)}\gamma_{i,n})^{j-1}\gamma_{1,i-1}^{(b_1,i-1)}\gamma_i^1, \dagger_{\bar{i}}, \langle_{\bar{i}}F\rangle_{\bar{i}}) \sim$

$\quad (\langle_{\bar{i}}F\rangle_{\bar{i}}, \dagger_{\bar{i}}, \gamma_{i+1,n}(\gamma_{1,i-1}^{(b_1,i-1)}\gamma_{i,n})^{m-j}\bar{\alpha}_{i,n+1}\alpha_{i,n+1}) \in R_{\mathsf{del}}$ where $b_i \in$
$\{0,1\}, 1 \le j \le m, 1 \le i \le n, \gamma_{1,1} = \gamma_1, \gamma_{n,n} = \gamma_n, \gamma_{1,0} = \gamma_{n+1,n} = \lambda.$
The context says that the substring $\dagger_{\bar{i}}\langle_{\bar{i}}F\rangle_{\bar{i}}$, which is going to be deleted, has to be preceded by a substring which is bordered by f_{i-1} in the left side, and followed by a substring of the form $f_{n+1} \cdots f_i f_i \cdots f_{n+1}$. One more crucial constraint is that all $\gamma_j, j < i$, have to be broken already and $\gamma_j, j > i$, have not been processed up to now. $\mathsf{del}_{\dagger}$ applies to m copies of Γ attached to δ, namely, it applies to all Γ in the maximally parallel way on the in-string which satisfies the contexts.

$\mathsf{del}_{\dagger_i} : (\bar{\alpha}_{i,n+1}\alpha_{i,n+1}\delta(\gamma_{1,i-1}^{(b_1,i-1)}\gamma_{i,n})^{j-1}\gamma_{1,i-1}^{(b_1,i-1)}, \dagger_i, \langle_i T\rangle_i) \sim$

$(\langle_i T\rangle_i, \dagger_i, \gamma_i^0\gamma_{i+1,n}(\gamma_{1,i-1}^{(b_1,i-1)}\gamma_{i,n})^{m-j}\bar{\alpha}_{i,n+1}f_{i-1}) \in R_{\mathsf{del}},$

where $b_i \in \{0,1\}, 1 \le j \le m, 1 \le i \le n, \gamma_{1,1} = \gamma_1,$

$\gamma_{n,n} = \gamma_n, \gamma_{1,0} = \lambda, \gamma_{n+1,n} = \lambda.$

A substring $\dagger_i\langle_i T\rangle_i$ with its preceding substring bordered by f_{i-1} in the right side is deleted by $\mathsf{del}_{\dagger_i}$.

Note that since there is no splicing relation for the repeats of $\dagger$ in the classes of the splicing schemes except R_{del}, only $\mathsf{del}_{\dagger}$ applies to the repeats of $\dagger$. Up to now, 2^i truth-assignments of the form $\gamma_1^{(b_1)}\gamma_2^{(b_2)}\ldots\gamma_i^{(b_i)}\gamma_{i+1}\ldots\gamma_n, b_i \in \{0,1\}$ have been generated on the in-string.

(3) At the $3k-1$th ($1 \le k \le n$) step of the computation, the deletion operation del_{f_i} is allowed. It applies for the repeats of f_i and deletes one copy of f_i with one f_{i-1} if it is available between those f_i.

$$\alpha f_i f_{i-1} f_i \alpha' \Longrightarrow_{\mathsf{del}_{f_i}}^{max} \alpha f_i \alpha', \alpha, \alpha' \in \Sigma^*, 1 \le i \le n.$$

The next two contexts are in the splicing scheme for del_{f_i}:

- $(\bar{\alpha}_{i+1,n+1}, f_i, f_{i-1}) \sim (f_{i-1}, f_i, \alpha_{i+1,n+1}\delta(\gamma_{1,i}^{b_1,i}\gamma_{i+1,n})^m) \in R_{\mathsf{del}},$
- $(\delta(\gamma_{1,i}^{(b_1,i)}\gamma_{i+1,n})^m\bar{\alpha}_{i+1,n+1}, f_i, f_{i-1}) \sim (f_{i-1}, f_i, \alpha_{i+1,n+1}) \in R_{\mathsf{del}}.$

Remember that at the previous two steps del_{f_i} was not applicable because γ_i had not been operated. Since γ_i was operated at the previous step by $\mathsf{del}_{\dagger_i}$, now del_{f_i} can be applied. Here the subscripts are shifted from i to $i-1$.

$$\alpha f_i f_i \alpha' \Longrightarrow_{\mathsf{del}_{f_i}}^{max} \alpha f_i \alpha', \text{ where}$$

- $(\delta(\gamma_{1,i}^{(b_1,i)}\gamma_{i+1,n})^m\bar{\alpha}_{i+1,n+1}, f_i, f_i) \sim (f_i, f_i, \alpha_{i+1,n+1}) \in R_{\mathsf{del}}$
  where $\alpha_{n+1,n+1} = \bar{\alpha}_{n+1,n+1} = f_{n+1}.$

After n repeats of the cyclic iteration (1) $\Longrightarrow$ (2) $\Longrightarrow$ (3), it ends at the $3n$th step and all truth-assignments have been generated completely. Now the in-string is of the form:

$$\pi_{3n} = \mathsf{YE}(f_{n+1}f_nf_{n+1}\delta\Gamma^m f_{n+1}f_nf_{n+1})^{2^n}\mathsf{ES}.$$

We enter to the next phase as follows.

(ii) *Assigning the truth-values to the variables.* The truth-values (truth-assignments) are assigned to each variable (to each clause) of the propositional formula by the next translocation operations:

$$\alpha'\langle_{\sigma(i_b)}x_{\sigma(i_b)}\rangle_{\sigma(i_b)}\alpha''\langle_{\sigma(i_b)}B\rangle_{\sigma(i_b)}\alpha''' \Longrightarrow_{\mathsf{trl}_{\langle_{\sigma(i_b)},\rangle_{\sigma(i_b)}}}^{max}$$

$$\alpha'\langle_{\sigma(i_b)}B\rangle_{\sigma(i_b)}\alpha''\langle_{\sigma(i_b)}x_{\sigma(i_b)}\rangle_{\sigma(i_b)}\alpha''',$$

where $b \in \{0,1\}, \sigma(i_1) = i, \sigma(i_0) = \bar{i}, 1 \le i \le n, \alpha', \alpha'', \alpha''' \in \Sigma^*$. We have the next two splicing relations in R_{trl}.

- $(\vee, \langle_{\sigma(i_b)}, x_{\sigma(i_b)}\rangle_{\sigma(i_b)} \vee u\dagger_{\sigma(i_b)}) \sim (x_{\sigma(i_b)}\rangle_{\sigma(i_b)} \vee u\dagger_{\sigma(i_b)}, \langle_{\sigma(i_b)}, B\rangle_{\sigma(i_b)}\dagger_{\sigma(i_b)} v)$
- $(\langle_{\sigma(i_b)} x_{\sigma(i_b)}, \rangle_{\sigma(i_b)}, \vee u\dagger_{\sigma(i_b)} \langle_{\sigma(i_b)} B) \sim (\vee u\dagger_{\sigma(i_b)} \langle_{\sigma(i_b)} B, \rangle_{\sigma(i_b)}, \dagger_{\sigma(i_b)} v)$,
    where $u \neq u' f_{n+1} u''$, $v \neq v' f_{n+1} v''$, $u', u'', v', v'' \in \Sigma^*$.

By the application of $\text{trl}_{\langle i,\rangle_i}$, every variable x_i flanked by $\langle_i$ and $\rangle_i$ in each clause c_j and the corresponding truth-value T (*true*) flanked by $\langle_i$ and $\rangle_i$ is swapped if such correspondence exists. Similarly, for assigning the truth-value F (*false*) to $x_{\bar{i}}$, the contents of $\langle_{\bar{i}} x_{\bar{i}}\rangle_{\bar{i}}$ and $\langle_{\bar{i}} F\rangle_{\bar{i}}$ are swapped by $\text{trl}_{\langle_{\bar{i}},\rangle_{\bar{i}}}$. Thus, all truth-assignments are assigned to their attached propositional formula at the same time in a single step. The truth-values assigned to a fixed formula should not be mixed from the different assignments. That is why the constraints $u \neq u' f_{n+1} u''$, $v \neq v' f_{n+1} v''$ are required in the context. It is the $3n + 1$th step.

(iii) Checking the satisfiability of the propositional formula. If a clause c_k is satisfied by a truth-assignment, then at least one substring of type $\langle_{\sigma(i_b)} B\rangle_{\sigma(i_b)}$ exists in c_k as $\$_k x_k \langle_{\sigma(i_b)} B\rangle_{\sigma(i_b)} y_k \$_k$. However, if the propositional formula φ is satisfied by a truth-assignment, then each clause c_l of φ is of the form $\$_l x_l \langle_{\sigma(i_b)} B\rangle_{\sigma(i_b)} y_l \$_l$, $x_l, y_l \in \Sigma^*, 1 \leq l \leq m$. Then an encoding of the propositional formula satisfied by an assignment is of the form:

$$\$_0 \$_1 x_1 \langle_{\sigma(i_b)} B\rangle_{\sigma(i_b)} y_1 \$_1 \$_2 x_2 \langle_{\sigma(i_b)} B\rangle_{\sigma(i_b)} y_2 \$_2 \ldots \$_m x_m \langle_{\sigma(i_b)} B\rangle_{\sigma(i_b)} y_m \$_m \$_{m+1},$$

where $\sigma(i_1) = i, \sigma(i_0) = \bar{i}, 1 \leq i \leq n, x_j, y_j \in \Sigma^*, 1 \leq j \leq m$.

The deletion operation $\text{del}_{\$_i}$ applies to the in-string in maximally parallel, and deletes the clauses which contain at least a truth-value of the form $\$_i \alpha' \langle_{\sigma(j_b)} B\rangle_{\sigma(j_b)} \alpha'' \$_i$ as follows:

$$\alpha \$_i \alpha' \langle_{\sigma(j_b)} B\rangle_{\sigma(j_b)} \alpha'' \$_i \alpha''' \Longrightarrow_{\text{del}_{\$_i}}^{max} \alpha \$_i \alpha''',$$

- $(\$_{i-1}, \$_i, x_i \langle_{\sigma(j_b)} B\rangle_{\sigma(j_b)} y_i) \sim (x_i \langle_{\sigma(j_b)} B\rangle_{\sigma(j_b)} y_i, \$_i, \$_{i+1}) \in R_{\text{del}}$,
    $x_i, y_i \in \Sigma^*, 1 \leq i \leq m, 1 \leq j \leq n$.

The checking phase has been done at the $3n + 2$th step of the computation.

(iv) Deciding. At the end of the computation, we obtain some sequences of the form $\$_0 \$_1 \$_2 \ldots \$_m \$_{m+1}$ on the in-string if there exist truth-assignments which satisfy the formula φ. Then del_E applies to the in-string $\text{YE} u \$_1 \$_2 \ldots \$_{m-1} \$_m v \text{ES}$ for the repeats of E.

$$\text{YE} u \$_1 \$_2 \ldots \$_{m-1} \$_m v \text{ES} \Longrightarrow_{\text{del}_E} \text{YES}, \text{ where}$$

- $(\text{Y}, \text{E}, \$) \sim (\$, \text{E}, \text{S}) \in R_{\text{del}}$, $\$ = u \$_1 \$_2 \ldots \$_{m-1} \$_m v$, for some $u, v \in \Sigma^*$.

Thus, at $3n + 3$th step, the target string YES is reached.

If no sequence $\$_1 \$_2 \ldots \$_m$ is obtained with the in-string, then the computation just halts at the $3n + 2$th step since no rule is possible to apply from now on, hence, π_0 is not accepted by G. Thus, the problem φ is solved in a linear time. $\square$

We recall here the following result of [6].

Theorem 2. *[6] For any deterministic Turing machine $M = (S, \Sigma \cup \{\#\}, P)$ there exists an intramolecular recombination system $G_M = (\Sigma', \sim, \alpha_0, w_t)$ and a string $\pi_M \in \Sigma'^*$ such that for any word w over Σ^* there exists $k_w \geq 1$ such that $w \in L(M)$ if and only if $w \#^5 \pi_M^{k_w} \#^2 \in L(G_M)$.*

The next theorem proves that eAIR system is computationally universal.

Theorem 3. *For any deterministic Turing machine* $M = (S, \Sigma \cup \{\#\}, P)$, *there exists an extended intramolecular recombination system* $G_M = (\Sigma', \sim, \widetilde{R}, \alpha_0, w_t)$, *and a string* $\pi_M \in \Sigma'^*$ *for any word* $w \in L(M)$ *iff* $w\#^3\pi_M\#^2 \in L(G_M)$.

Proof. Since the string can be extended by cpy during the recombination processes in eAIR, with this proof we do not need to reserve as many copies of an encoded TM rule as we did in Theorem 2. Instead, an encoding of a rewriting rule is duplicated by cpy before it is used. Then one copy of the encoding is used by a translocation operation while another copy is reserved as initial state. In a manner similar to Theorem 2, for a Turing machine M we construct an extended intramolecular recombination system

$$G_M = (\Sigma', \sim, \widetilde{R}, \alpha_0, w_t), \text{ where}$$
$$\Sigma' = S \cup \Sigma \cup \{\#\} \cup \{\$_i \mid 0 \leq i \leq m+1\},$$
$$\alpha_0 = \#^3 s_0, w_t = \#^3 s_f \#^3,$$
$$\widetilde{R} = \{\text{trl}_{p,q}, \text{del}_p, \text{cpy}_p \mid p, q \in core(R)\}.$$

We also consider the string

$$\pi_M = \$_0 \Big(\prod_{\substack{1 \leq i \leq m \\ p,q \in \Sigma \cup \{\#\}}} \$_i p v_i q \$_i \Big) \$_{m+1}, \ \pi_M \in \Sigma'^*.$$

The classes of splicing relations $R_{op \in \widetilde{R}} = (\Sigma', \sim, op)$ are constructed as follows:

$$R_{\text{cpy}}: \ (p u_i q w' \$_0 w'', \$_i, p v_i q) \sim (p v_i q, \$_i, \$_{i+1}), \tag{1}$$
$$R_{\text{trl}}: \ (c, p, u_i q d) \sim (\$_i, p, v_i q p), \tag{2}$$
$$(p u_i, q, d) \sim (\$_i p v_i, q, p v_i q \$_i), \tag{3}$$
$$\text{where } c, d, p, q \in \Sigma \cup \{\#\},$$
$$R_{\text{del}}: \ (\$_i, p, u_i q) \sim (u_i q, p, v_i q \$_i), \tag{4}$$
$$(\#^3 s_f \#, \#, \#) \sim (\$_{m+1}, \#, \#). \tag{5}$$

If a word $w \in \Sigma^*$ is accepted by Turing machine M, then associated eAIR G_M works as follows:

$$\#^3 s_0 w \#^3 \pi_M \#^2 \Longrightarrow_{R_{op \in \widetilde{R}}}^* \#^3 s_f \#^3 \pi_M \#^2 \Longrightarrow_{\text{del}_\#} \#^3 s_f \#^3.$$

Where we refer to the subsequence $\#^3 s_0 w \#^3$ as the "data", and to the subsequence $\pi_M \#^2$ as the "program". A rewriting rule $i : u \to v \in P$ of M is simulated in three subsequent steps (cpy $\Longrightarrow$ trl $\Longrightarrow$ del) in G_M.

Step 1. When left-hand side of a rule $i : u_i \to v_i$ appears in the form $p u_i q$ in data, the corresponding encoding of the right-hand side of the rule $p v_i q$ flanked by $\$_i$-s in the program is copied ($\$_i p v_i q p v_i q \$_i$). The copy cpy$_{\$_i}$ performs in the contexts defined by relation (1) as follows,

$$x p u_i q w' \$_0 w'' \$_i p v_i q \$_i \$_{i+1} y \Longrightarrow_{\text{cpy}_{\$_i}} x p u_i q w' \$_0 w'' \$_i p v_i q p v_i q \$_i \$_{i+1} y.$$

Step 2. The u_i and the corresponding v_i each one is flanked by p and q are swapped by $\mathsf{trl}_{p,q}$ provided the relations (2) and (3) are satisfied. The translocation $\mathsf{trl}_{p,q}$ performs as follows,

$$xcpu_iqdw'\$_ipv_iqpv_iq\$_i\$_{i+1}y \Longrightarrow_{\mathsf{trl}_{p,q}} xcpv_iqdw'\$_ipu_iqpv_iq\$_i\$_{i+1}y.$$

Step 3. The substring pu_iq used in the swapping is deleted from the program in the context of relation (4) as follows,

$$xcpu_jqdw'\$_ipu_iqpv_iq\$_i\$_{i+1}y \Longrightarrow_{\mathsf{del}_p} xcpu_jqdw'\$_ipv_iq\$_i\$_{i+1}y.$$

The encoding of $\$_ipv_iq\$_i$ is still kept in program if u_i appears again in the data.

Thus, the rewriting rule $i : u_i \to v_i$ is simulated in three subsequent *Steps 1–3* in G_M. At step 3, while del_p performs, the next rewriting rule $j : u_j \to v_j$, $j \neq i$ simulation could start making a copy of the corresponding $\$_jpv_jq\$_j$ in program, unless u_i appears immediately in a subsequent step in data. If the last case happens, its simulation starts at the next step following del_p.

The rewriting rules of M are simulated in G_M correctly by the subsequent repeats of the recombination *Steps 1–3*, and the next string can be reached:

$$\#^3 s_0 w \#^3 \pi_M \#^2 \Longrightarrow^*_{R_{op \in R}} \#^3 s_f \#^3 \pi_M \#^2.$$

When the substring $\#^3 s_f \#^3$ contains the final state s_f is obtained in data, the contexts of the relation (5) are satisfied the operation $\mathsf{del}_\#$ to be applied:

$$\#^3 s_f \#^3 \pi_M \#^2 \Longrightarrow_{\mathsf{del}_\#} \#\#\# s_f \#\#\#.$$

Thus, the target $\#\#\# s_f \#\#\#$ is reached, G_M accepts the word $\#^3 s_0 w \#^3 \pi_M \#^2$.

For the converse implication, we have to claim that only a correct relation or a correct combination of the relations and nothing else is used in each recombination step. We start the claim with the relation (1): operation $\mathsf{cpy}_{\$_i}$ does not repeat immediately for the repeats of the pointer $\$_i$ following its application. By the definition of cpy_p, the substring flanked by the repeat of a pointer p is copied if and only if it is the following context of the first occurrence of p, and the preceding context of the second occurrence of p in the splicing relation (see Definition 3). Since the substring $\$_ipv_iqpv_iq\$_i$ does not satisfy the context defined by relation (1), $\mathsf{cpy}_{\$_i}$ is not applicable to it. It is true that only relations (2) and (3) from R_{trl} among others are satisfied on the string at the next step of $\mathsf{cpy}_{\$_i}$ applied. When $\mathsf{trl}_{p,q}$ was applied at *Step 2* swapping u_i and v_i in the contexts (2) and (3), it cannot be applied immediately its following step to the same pointers in the same contexts as previous one. For instance, encoding of u_i would appear again in the data by $\mathsf{trl}_{p,q}$ as $xpu_iqw\$_ipu_iqpv_iq\$_i$, but $\mathsf{trl}_{p,q}$ is not applicable to p and q here. Because the preceding and following contexts of the second occurrence of the pointer q of the splicing relation (3) cannot be satisfied on the string since u_i and v_i can never be the same strings in the rules of Turing machine M. It completes the proof.   □

5   Final Remarks

In the present paper we propose a computing model which is based on the ciliate intramolecular gene assembly model developed, for instance, in [1]. The model is mathematically elegant and biologically well-motivated because only three types of operations (two of them are formalizations of gene assembly process in ciliates) and a single string are involved. The context-sensitivity for string operations are already well-known in formal language theory, see [9,7]. Moreover, parallelism is a feature characteristic of bio-inspired computing models, starting with DNA computing, which is also the case with our model. From a computer science point of view, in the model (eAIR system) is both as powerful as Turing machines and as efficient in solving intractable problems in feasible time. Investigating the other computability characteristics of eAIR system could be worthwhile.

It is important to note that the cpy operation we introduce in this model is purely theoretical. Although duplication mechanisms exists in nature, implementing a copy operation of the kind we consider has not been demonstrated yet.

Acknowledgments. The work of T.-O.I. is supported by the Center for International Mobility (CIMO) Finland, grant TM-06-4036 and by Academy of Finland, project 203667. The work of I.P. is supported by Academy of Finland, project 108421.

We are grateful to Gheorghe Păun for useful discussions.

References

1. Ehrenfeucht, A., Harju, T., Petre, I., Prescott, D.M., Rozenberg, G.: Computation in Living Cells: Gene Assembly in Ciliates. Springer, Heidelberg (2003)
2. Galiukschov, B.S.: Semicontextual grammars, Mathematika Logica i Matematika Linguistika, Talinin University, pp. 38–50, 1981 (in Russian)
3. Garey, M., Jonhson, D.: Computers and Interactability. A Guide to the Theory of NP-completeness. Freeman, San Francisco, CA (1979)
4. Harju, T., Petre, I., Li, C., Rozenberg, G.: Parallelism in gene assembly. In: Proceedings of DNA-based computers 10, Springer, Heidelberg, 2005 (to appear)
5. Head, T.: Formal Language Theory and DNA: an analysis of the generative capacity of specific recombinant behaviors. Bull. Math. Biology 49, 737–759 (1987)
6. Ishdorj, T.-O., Petre, I., Rogojin, V.: Computational Power of Intramolecular Gene Assembly. International Journal of Foundations of Computer Science (to appear)
7. Kari, L., Thierrin, G.: Contextual insertion/deletions and computability. Information and Computation 131, 47–61 (1996)
8. Landweber, L.F., Kari, L.: The evolution of cellular computing: Nature's solution to a computational problem. In: Proceedings of the 4th DIMACS Meeting on DNA-Based Computers, Philadelphia, PA, pp. 3–15 (1998)
9. Marcus, S.: Contextual grammars. Revue Roumaine de Matématique Pures et Appliquées 14, 1525–1534 (1969)
10. Papadimitriou, Ch.P.: Computational Complexity. Addison-Wesley, Reading, MA (1994)

11. Păun, Gh.: Marcus Contextual Grammars. Kluwer, Dordrecht (1997)
12. Păun, Gh., Rozenberg, G., Salomaa, A.: DNA Computing - New computing paradigms. Springer, Heidelberg (1998)
13. Salomaa, A.: Formal Languages. Academic Press, New York (1973)
14. Searls, D.B.: Formal language theory and biological macromolecules. Series in Discrete Mathematics and Theoretical Computer Science 47, 117–140 (1999)

Learning Vector Quantization Network for PAPR Reduction in Orthogonal Frequency Division Multiplexing Systems

Seema Khalid[1], Syed Ismail Shah[2], and Jamil Ahmad[2]

[1] Center for Advanced Studies in Engineering, Islamabad, Pakistan
[2] Iqra University Islamabad Campus H-9, Islamabad, Pakistan
k_seema6@yahoo.com, {ismail,jamilahmad}@iqraisb.edu.pk

Abstract. Major drawback of Orthogonal Frequency Division Multiplexing (OFDM) is its high Peak to Average Power Ratio (PAPR) that exhibits inter modulation noise when the signal has to be amplified with a non linear high power amplifier (HPA). This paper proposes an efficient PAPR reduction technique by taking the benefit of the classification capability of Learning Vector Quantization (LVQ) network. The symbols are classified in different classes and are multiplied by different phase sequences; to achieve minimum PAPR before they are transmitted. By this technique a significant reduction in number of computations is achieved.

Keywords: Orthogonal Frequency Division Multiplexing, Peak to Average Power Ratio, Learning Vector Quantization, Neural Network.

1 Introduction

Orthogonal Frequency Division Multiplexing (OFDM) has become a popular modulation method in high-speed wireless communications. By partitioning a wideband fading channel into narrowband channels, OFDM is able to mitigate the detrimental effects of multi-path fading using a simple one-tap equalizer. However, in the time domain OFDM signals suffer from large envelope variations, which are often characterized by the Peak-to-Average Power ratio (PAPR). The PAPR of the signal x(t) with instantaneous power P(t) is defined as:

$$\mathrm{PAPR}\big[x(t)\big] = \frac{\displaystyle\max_{0 \le t \le T_s} \big|x(t)\big|^2}{\dfrac{1}{T_s} \displaystyle\int_0^{T_s} \big|x(t)\big|^2 \, dt} = \frac{\displaystyle\max_{0 \le t \le Ts} \big[P(t)\big]}{E\{P(t)\}} \tag{1}$$

Due to high PAPR, signals require that transmission amplifiers operate at very low power efficiencies to avoid clipping and distortion. Different methods have been proposed during the last decade in order to reduce high PAPR of the OFDM modulation. Those might be put into two big groups, linear techniques and non-linear techniques. Some examples of methods using linear techniques are the Peak

S.G. Akl et al.(Eds.): UC 2007, LNCS 4618, pp. 106–114, 2007.

Reduction Carriers (PRC) schemes [1], Selected Mapping (SLM) [2], Partial transmit Sequences (PTS) [2] and block code. On the other hand, non-linear techniques modify the envelope of the time domain signal and are mainly composed of clipping and windowing schemes [3], peak canceling Schemes [4] and Companding technique based methods [5]. Along with conventional techniques, attempts have been made to reduce PAPR of OFDM signal using unconventional computing as reported in [6] [7] [8].

Usefulness of any proposed scheme is that it reduces the PAPR and is not computationally complex. In our work we modified an existing promising reduction technique, SLM by using Learning Vector Quantization (LVQ) network. In a SLM system, an OFDM symbol is mapped to a set of quasi-independent equivalent symbols and then the lowest-PAPR symbol is selected for transmission. The tradeoff for PAPR reduction in SLM is computational complexity as each mapping requires an additional Inverse Fast Fourier transform (IFFT) operation in the transmitter. By the Learning Vector Quantization based SLM (LVQ-SLM), we eliminate the additional IFFT operations, getting a very efficient PAPR reduction technique with reduced computational complexity.

The rest of the paper is organized as follows. A brief overview of SLM technique is given in Section 2. Section 3 describes the LVQ algorithm. The proposed model is explained in Section 4. Results are reported in Section 5 followed by conclusion in Section 6.

2   Selected Mapping Technique for PAPR Reduction

SLM takes advantage of the fact that the PAPR of an OFDM signal is very sensitive to phase shifts in the frequency-domain data. PAPR reduction is achieved by multiplying independent phase sequences to the original data and determining the PAPR of each phase sequence/data combination. The combination with the lowest PAPR is transmitted. In other words, the data sequence X is element-wise phased by D, N-length phase sequences, $\left\{\varphi[k]^{(d)}\right\}_{k=0}^{N-1} = \varphi^{(d)}$ where d is an integer such that d ∈ [0; D- 1] and D is total number of different phase sequences. After phasing, the D candidate frequency-domain sequences are given by

$$X^{(d)} = X \circ e^{j\varphi^{(d)}} \qquad (2)$$

Where '∘' is element-wise multiplication. We assume that $\varphi^{(0)} = 0 \,\forall k$ so that $X^{(0)} = X$ .Define the D candidate time-domain OFDM symbols $x_L^{(d)} = \text{IFFT}_L\left\{X^{(d)}\right\}$. Note that all of the candidate symbols carry the same information. In order to achieve a PAPR reduction, the symbol with the lowest PAPR is transmitted. We define

$$d = \underset{0<d<D}{\arg\ \min}\text{PAPR}\left\{x_L^{(d)}\right\} \qquad (3)$$

With d , transmitted signal is $x^{(d)}$. At the receiver the X can be recovered with de-rotating i.e.

$$X = FFT\{x^{(d)}\} \circ e^{-j\varphi^{(d)}} \tag{4}$$

To recover X it is necessary for the receiver to have a table of all phase sequences $\varphi^{(d)}$. A $\lceil \log_2 D \rceil$ bit of side information is required to recover original symbol.

3 Learning Vector Quantization

LVQ is a supervised version of vector quantization, similar to Self Organizing Maps (SOM) [9], [10]. It is used in pattern recognition, multi-class classification and data compression tasks. LVQ algorithms directly define class boundaries based on prototypes, a nearest-neighbor rule and a winner-takes-it-all paradigm. The main idea is to cover the input space of samples with 'code vectors' (CVs), each representing a region labeled with a class. A CV can be seen as a prototype of a class member, localized in the centre of a class or decision region ('Voronoi cell') in the input space. As a result, the space is partitioned by a 'Voronoi net' of hyper planes perpendicular to the linking line of two CVs shown in Fig. 1. A class can be represented by an arbitrarily number of CVs, but one CV represents one class only. LVQ networks can classify any set of input vectors, not just linearly separable sets of input vectors. The only requirement is that the competitive layer must have enough neurons, and each class must be assigned enough competitive neurons [11]. This property of LVQ network is used to segregate the symbol that results in high PAPR from those which result in low PAPR.

It has been shown that LVQ is particularly well suited as an on-line classifier primarily due to its algorithmic speed [12]. The idea of LVQ is that in a classification system based on the nearest-neighbor rule, a drastic gain of computation speed can be obtained reducing the number of vectors that represent each class. A set of reference vectors, also called code vectors (CVs), is adapted through an iterative process to the

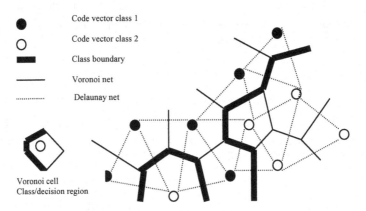

Fig. 1. Tessellation of input space into decision/class regions by code vectors represented as neurons positioned in two-dimensional feature space

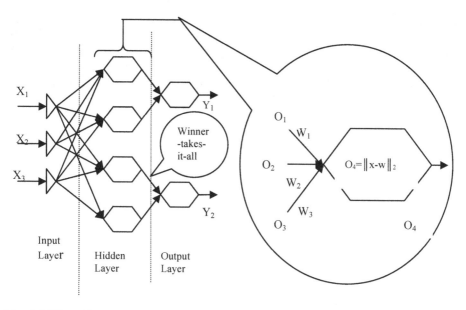

Fig. 2. LVQ architecture: adjustable weight between input and hidden layer and a winner takes it all mechanism

data according to a competitive learning rule. Learning means modifying the weights in accordance with adapting rules and, therefore, changing the position of a CV in the input space.

3.1 LVQ Algorithm

If $x(t)$ the training vector presented at iteration t, m_j the set of code vectors and $m_c(t)$ the nearest code vector to $x(t)$. Vector m_c is obtained from the equation:

$$\| x(t)-m_c(t) \| = \min \| x(t)-m_j(t) \| \qquad (5)$$

And is adapted according to the learning rule

$$m_c(t+1) = m_c(t)+ \alpha(t)\cdot[x(t)-m_c(t)] \qquad \text{if } C(x) =C(m_c)$$

or

$$m_c(t+1)=m_c(t)- \alpha(t)\,[x(t)-m_c(t)] \qquad \text{if } C(x) \neq C(m_c). \qquad (6)$$

Where $\alpha(t) \in [0,1]$ is a learning rate and $C(\cdot)$ is a function returning the class of a vector. The result of the process is an approximation of the data probability density function by the code vectors. After such a training of the code vectors, any vector 'y' is classified according to the nearest neighbor rule. The 'y' is classified into class C_k if the nearest code vector to 'y' is m_c, where $C(m_c) =C_k$.

4 Proposed Model

We tabulate all the possible symbols for a Binary Phase Shift Keyed (BPSK) and 8 sub-carrier OFDM system. There are 256 possible symbols. After taking their IFFT, we compute the PAPRs of these symbols and segregate the symbols into two classes. Those with high PAPR (> 4) are affirmed in class A and those which results in low PAPR, in class B. A LVQ classifier is off-line trained to classify the input vectors in two predefined classes. The trained network classifies the incoming symbols and symbols classified in class A are rotated with a predefined phase value. One bit side information is required to be transmitted to tell the receiver that, it's an original or rotated symbol. A simple rotator is used in the receiver that rotates (with the same phase) only those symbols which belong from class A in opposite direction.

Then we take three different phase sequences. We tabulate all the PAPRs of original symbols and those of rotated symbols with different phase values. We classified the input data (256 symbols) into four different classes. Class O contains symbols without phase shift, which results in minimum PAPRs as compared to rest of their three rotated versions. Class A comprised of those symbols which result in

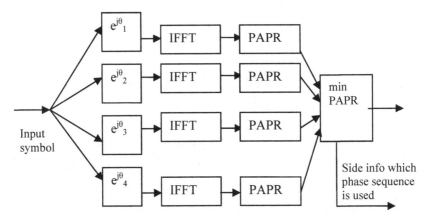

Fig. 3. The block diagrams of the SLM technique

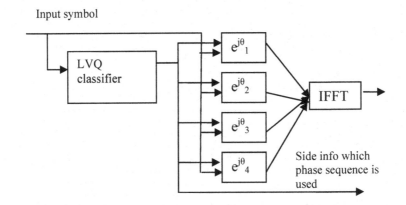

Fig. 4. The block diagrams of LVQ-SLM technique

minimum PAPR when rotated with phase sequence A. Similarly class B and class C are defined with symbols having minimum PAPR after being phase shifted by sequence B or sequence C. All of these computations are done as pre-processing and the input symbols are distributed in four different classes O, A, B and C. A LVQ classifier is trained offline to classify the input symbols. Two bit side information is required to identify to which class the symbol belonged. At the receiver side three phase rotator are used to de-rotate the symbol depending, from which class the symbol belonged. Fig. 3 shows the block diagrams of the SLM technique. Our proposed model of SLM with a LVQ classifier is shown in Fig. 4. The receivers are identical for both the techniques.

5. Simulation Results

The accuracy of classification depends on several factors. A learning schedule (a plan for LVQ-algorithms LVQ1 or LVQ2) with specific values for the main parameters at different training phases and the number of CVs for each class is decided while avoiding under- or over fitting. Additionally, the rule for stopping the learning process as well as the initialization method determined the results.

The simulation results in which Complementary Cumulative Distribution Function (CCDF) of PAPR are shown in Fig. 5. D=1 is when no classification is done. D=2 is the case when symbols are classified in two classes, higher valued (PAPR>4) in class A and lower valued in class B, and class A symbols are multiplied by a fixed phase sequence. D=4 is when we segregate the data in four classes and symbol is multiplied by the predefined phase sequence to have minimum PAPR among the four mappings.

The simulations are repeated till convergence using different learning rates and varying the number of CVs. Training is done offline so we kept learning rate small to have better convergence. The basic aim was to achieve better classification accuracy

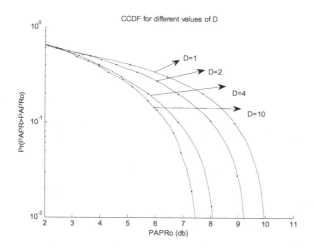

Fig. 5. CCDF of 8 sub-carrier BPSK OFDM signal With D=1, 2, 4, 10

with least number of CVs to avoid over fitting. Several sets of initial weight vectors are tested, and the best are selected.

It is observed in Fig. 5. (for D=2), the PAPR do not reduce much although the symbols with higher PAPR are segregated in a different class and multiplied with a phase sequence. The reason is a single phase sequence is used so it does not ensure reduced PAPR after phase rotation. All the more an LVQ classifier is not 100% accurate and miss classified some symbols. In case of three different mapping (D=4) and then choose the symbol having minimum PAPR we get better results. The classification accuracy is not improved but the probability that the symbol is wrongly classified and at the same time has high PAPR was very less. By this model we achieved as much reduction in PAPR as from the conventional SLM scheme (Fig. 6.) but the computational cost has reduced significantly.

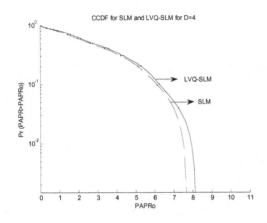

Fig. 6. CCDF of SLM and LVQ-SLM with D=4 for 8-BPSK OFDM signal

For N sub carriers (N point IFFT) BPSK and D different phase sequences the comparison of computations involved are shown in table 1.

Table 1. Reduction in number of computations

Conventional SLM	Proposed model of LVQ-SLM
D IFFT operations (one N point IFFT costs $N/2\log_2(N)$ complex multiplications $N\log_2(N)$ complex addition)	Single IFFT operation
(D-1) PAPR calculations N length maximum value search	No PAPR to be calculated
Finding minimum from PAPR values. D length minimum value search	No minima is to be found

We get rid of online multiple IFFT and PAPR evaluating computations on the transmitter end where each IFFT cost many elementary operations. All the training of the LVQ classifier was done offline. We avoid over fitting (large number of CVs to define the class) to avoid unnecessary computations in online classification. The result of the training of classifier is tabulated in table 2. The table shows the percentages of data used for training and the percentage of errors occurred in classification. As the training is done offline so we used 100% data to get more accurate classifier.

Table 2. Percentage of training set and the error in classification

Training data	% error
100%	97
80%	88
50%	62

6 Conclusion

The proposed scheme, using classifier has achieved required PAPR reduction with significant saving in the number of computations. The underline idea is to use classification capability of LVQ network; classify the symbols in different classes depending on what set of operations the symbol require undergoing to result in low PAPR. Apply predefined set of operations to reduce PAPR of those symbols. Class identity, in which the symbol is classified at the transmitter end, is to be transmitted as side information. On the receiver, reverse set of operations are used to retrieve the original symbol, in view of its class information. By this idea of classifying the symbols in different groups, other nonlinear PAPR reduction schemes can be implemented with reduced computational cost.

Acknowledgement

The authors acknowledge the enabling role of the Higher Education Commission Islamabad, Pakistan and appreciate its financial support through "Development of S&T Manpower through Merit Scholarship scheme 200."

Reference

1. Tan, C.E., Wassell, I.J.: Suboptimum peak reduction carriers for OFDM systems. In: Proceeding for Vehicular Technology conference VTC-2003, vol. 2, pp. 1183–1187 (2003)
2. Muller, S.H., Bauml, R.W., Fisher, R.F.H.: OFDM with reduced Peak to Average power ratio by multiple signal representation: Annals of telecommunications, vol. 52(1-2) (February 1997)
3. Nee, R.V., Prasad, R.: OFDM for wireless Communications: Ed Artech House, U.S.A (2000)

4. May, T., Rohling, H.: Reducing Peak to Average Power Ratio reduction in OFDM Radio transmission system. In: Proceedings of IEEE VTC'98, Ottawa, Canada, vol. 3, pp. 2474–2478 (May 1998)
5. Tang, W., et al.: Reduction in PAPR of OFDM system using revised Companding. In: Priss, U., Corbett, D.R., Angelova, G. (eds.) ICCS 2002. LNCS (LNAI), vol. 2393, pp. 62–66. Springer, Heidelberg (2002)
6. Ohta, M., Yamashita, K.: A Chaotic Neural Network for Reducing the Peak-to-Average Power Ratio of Multi-carrier Modulation. In: International Joint Conference on Neural Networks, pp. 864–868 (2003)
7. Yamashita, K., Ohta, M.: Reducing peak-to-average power ratio of multi-cannier modulation by Hopfield neural network. Electronics Letters, 1370–1371 (2002)
8. Ohta, M., Yamada, H., Yamashita, K.: A neural phase rotator for PAPR reduction of PCC-OFDM signal: IJCNN'05, vol. 4, pp. 2363–2366 (August 2005)
9. Linde, Y., Buzo, A., Gray, R.M.: An algorithm for Vector Quantizer design. IEEE Trans. on communications, 702–710 (1980)
10. Yin, H., Alinson, N.M.: Stochastic analysis and comparison of Kohonen SOM with optimal filter. In: Third International Conference on Artificial Neural Networks Volume, pp. 182–185 (1993)
11. Neural Network Toolbox Matlab 7 Documentation
12. Flotzinger, D., Kalcher, J., Pfurtscheller, G.: Suitability of Learning Vector Quantization for on-line learning. In: A case study of EEG classification: Proc WCNN'93 World Congress on Neural Networks, vol. I, pp. 224–227. Lawrence Erlbaum, Hillsdale, NJ (1993)

Binary Ant Colony Algorithm for Symbol Detection in a Spatial Multiplexing System

Adnan Khan[1], Sajid Bashir[1], Muhammad Naeem[2], Syed Ismail Shah[3], and Asrar Sheikh[4]

[1] Centre for Advanced Studies in Engineering, Islamabad, Pakistan
[2] Simnon Fraser University Burnaby BC Canada
[3] Iqra University Islamabad Campus, H-9, Islamabad, Pakistan
[4] Foundation University, Islamabad, Pakistan
adkhan100@gmail.com

Abstract. While an optimal Maximum Likelihood (ML) detection using an exhaustive search method is prohibitively complex, we show that binary Ant Colony Optimization (ACO) based Multi-Input Multi-Output (MIMO) detection algorithm gives near-optimal Bit Error Rate (BER) performance with reduced computational complexity. The simulation results suggest that the reported unconventional detector gives an acceptable performance complexity trade-off in comparison with conventional ML and non-linear Vertical Bell labs Layered Space Time (VBLAST) detectors. The proposed technique results in 7-dB enhanced BER performance with acceptable increase in computational complexity in comparison with VBLAST. The reported algorithm reduces the computer time requirement by as much as 94% over exhaustive search method with a reasonable BER performance.

Keywords: Spatial Multiplexing System, Binary Ant System, Symbol detection, Multi-Input Multi-Output–Orthogonal Frequency Division Multiplexing (MIMO-OFDM).

1 Introduction

Real world optimization problems are often so complex that finding the best solution becomes computationally infeasible. Therefore, an intelligent approach is to search for a reasonable approximate solution with lesser computational complexity. Many techniques have been proposed that imitate nature's own ingenious ways to explore optimal solutions for both single and multi-objective optimization problems. Earliest of the nature inspired techniques are genetic and other evolutionary heuristics that evoke Darwinian evolution principles.

Ant Colony Optimization (ACO) meta-heuristics is one such technique that is based on the cooperative forging strategy of real ants [1],[2]. In this approach, several artificial ants perform a sequence of operations iteratively. Ants are guided by a greedy heuristic algorithm which is problem dependent, that aid their search for better solutions iteratively. Ants seek solutions using information gathered previously to

S.G. Akl et al.(Eds.): UC 2007, LNCS 4618, pp. 115–126, 2007.
© Springer-Verlag Berlin Heidelberg 2007

perform their search in the vicinity of good solutions. In [3], Dorigo showed that ACO is suitable for NP-complete problems such as that of the traveling salesman. The resistance of ACO to being trapped in local minima and convergence to near optimal solution in fewer iterations makes it a suitable candidate for real-time NP-hard communication problems. Its binary version known as binary ant system (BAS) is well suited for constrained optimization problems with binary solution structure [4].

In wireless communications system significant performance gains are achievable when Multi-Input Multi-Output (MIMO) architecture employing multiple transmit and receive antennas is used [6]. Increased system capacity can be achieved in these systems due to efficient exploitation of spatial diversity available in MIMO channel. This architecture is particularly suitable for higher data rate multimedia communications [7].

One of the challenges in designing a wideband MIMO system is tremendous processing requirements at the receiver. MIMO symbol detection involves detecting symbol from a complex signal at the receiver. This detection process is considerably complex as compared to single antenna system. Several MIMO detection techniques have been proposed [8]. These detection techniques can be broadly divided into linear and non-linear detection methods. Linear methods offer low complexity with degraded BER performance as compared to non-linear methods. This paper focuses on non-linear detectors and makes an effort to improve BER performance at the cost of complexity and vice versa. ML and V-BLAST detectors [9],[10] are well known non-linear MIMO detection methods. ML outperforms VBLAST in BER performance, while VBLAST is lesser complex than ML. In [11],[12] a performance complexity trade off between the two methods have been reported.

Being NP-hard [8] computational complexity of optimum ML technique is generically exponential. Therefore, in order to solve these problems for any non-trivial problem size, exact, approximate or un-conventional techniques such as meta-heuristics based optimization approach can be used. The exact method exploits the structure of the lattice and generally obtains the solution faster than a straightforward exhaustive search [8]. Approximation algorithm provides approximate but easy to implement low-complexity solutions to the integer least-squares problem. Whereas, meta-heuristics based algorithm works reasonably well on many cases, but there is no proof that it always converges fast like evolutionary techniques.

Earlier, one such evolutionary meta-heuristics known as particle swarm optimization (PSO) has been successfully applied to the MIMO detection for the first time by the authors [5]. In this paper, we report a binary ant system based symbol detection algorithm (BA-MIMO) for spatial multiplexing systems with an acceptable performance complexity trade off.

The rest of the paper is organized as follows. Section 2 provides the system model. In section 3 MIMO symbol detection problem for flat fading channel is described. A brief overview of the existing MIMO detectors is given in section 4. Section 5 provides the details of the proposed BA-MIMO detection technique. Performance of the proposed detector is reported in section 6, while section 7 concludes the paper.

2 MIMO Channel Model

Consider a MIMO system where N_t different signals are transmitted and arrive at an array of N_r $(N_t \le N_r)$ receivers via a rich-scattering flat-fading environment. Grouping all the transmitted and received signals into vectors, the system can be viewed as transmitting an N_t x 1 vector signal $\mathbf{x}$ through an N_t x N_r matrix channel $\mathbf{H}$, with N_r x 1 Gaussian noise vector $\mathbf{v}$ added at the input of the receiver. The received signal as an N_r x 1 vector can be written as:

$$\mathbf{y} = \mathbf{Hx} + \mathbf{v} \tag{1}$$

Where $\mathbf{y}$ is the received The $(n_r, n_t)^{\text{th}}$ element of $\mathbf{H}$, $h_{n_r n_t}$, is the complex channel response from the n_t^{th} transmit antenna to the n_r^{th} receive antenna. The transmitted symbol $\mathbf{x}$ is zero mean and has covariance matrix $\mathbf{R_x} = \mathrm{E}\{\mathbf{xx^*}\} = \sigma_x^2 \mathbf{I}$. The vector $\mathbf{v}$ is also zero-mean and $\mathbf{R_v} = \mathrm{E}\{\mathbf{vv^*}\} = \sigma_v^2 \mathbf{I}$. In frequency-selective fading channels, the entire channel frequency response $h_{n_r n_t}$ is a function of the frequency. We can therefore, write:

$$\mathbf{y}(f) = \mathbf{H}(f)\mathbf{x}(f) + \mathbf{v}(f) \tag{2}$$

When OFDM modulation is used, the entire channel is divided into a number of sub-channels. These sub-channels are spaced orthogonally to each other such that no inter-carrier interference (ICI) is present at the sub-carrier frequency subject to perfect sampling and carrier synchronization. When sampled at the sub-carrier frequency of f_{n_c}, the channel model becomes.

$$\mathbf{y}^{(n_C)} = \mathbf{H}^{(n_C)}\mathbf{x}^{(n_C)} + \mathbf{v}^{(n_C)}; \quad n_C = -N_C/2, \ldots, N_C/2-1 \tag{3}$$

With N_c, the number of subcarriers is sufficiently large, the sub-channel at each of these sub-carriers will experience flat-fading. Therefore, when using OFDM, the MIMO detection over frequency-selective channels is transformed into MIMO detection over N_c narrowband flat-fading channels. For this reason, we only focus on the MIMO detection algorithms in flat-fading channels. The entries of the channel matrix $\mathbf{H}$ are assumed to be known to the receiver but not to the transmitter. This assumption is reasonable if training or pilot signals are sent to estimate the channel, which is constant for some coherent interval.

3 Problem Formulation

The task is that of detecting N_t transmitted symbols from a set of N_r observed symbols that have passed a non-ideal communication channel, typically modeled as a linear system followed by an additive noise vector as shown in Fig. 1.

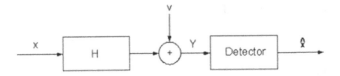

Fig. 1. A simplified linear MIMO communication system showing the following discrete signals: transmitted symbol vector $x \in \chi^{N_t}$, channel matrix $\mathbf{H} \in \mathbb{R}^{N_t x N_r}$, additive noise vector $\mathbf{v} \in \mathbb{R}^{N_t}$, receive vector $\mathbf{y} \in \mathbb{R}^{N_t}$, and detected symbol vector $\hat{x} \in \mathbb{R}^{N_r}$

Transmitted symbols from a known finite alphabet $\chi = \{x_1,...,x_M\}$ of size M are passed to the channel. The detector chooses one of the M^{N_t} possible transmitted symbol vectors from the available data. Assuming that the symbol vectors $x \in \chi^{N_t}$ are equiprobable, the *Maximum Likelihood (ML)* detector always returns an optimal solution according to the following:

$$x_* \triangleq \underset{x \in \chi^{N_t}}{\arg \max} \, P(y \, is \, observed | x \, was \, sent) \tag{4}$$

Assuming the additive noise $\mathbf{v}$ to be white and Gaussian, the ML detection problem of Figure 1 can be can be expressed as the minimization of the squared Euclidean distance to a target vector $\mathbf{y}$ over N_t-dimensional finite discrete search set:

$$x_* = \underset{x \in \chi^{N_t}}{\arg \min} \, \|y - Hx\|^2 \tag{5}$$

Optimal ML detection scheme needs to examine all M^{N_t} or 2^{bN_t} symbol combinations (b is the number of bits per symbol). The problem can be solved by enumerating over all possible $\mathbf{x}$ and finding the one that causes the minimum value as in (5). Therefore, the computational complexity increases exponentially with constellation size M and number of transmitters N_t.

The reported BA-MIMO detector views this MIMO-OFDM symbol detection issue as a combinatorial optimization problem and tries to approximate the near-optimal solution iteratively.

4   Existing MIMO Detectors

4.1   Linear Detectors

A straightforward approach to recover x from y is to use an $N_t \times N_r$ weight matrix $\mathbf{W}$ to linearly combine the elements of $\mathbf{y}$ to estimate $\mathbf{x}$, i.e. $\hat{x} = \mathbf{W}\mathbf{y}$.

4.1.1   Zero-Forcing(ZF)

The ZF algorithm attempts to null out the interference introduced from the matrix channel by directly inverting the channel with the weight matrix [8].

4.1.2 Minimum Mean Squared Error (MMSE)

A drawback of ZF is that nulling out the interference without considering the noise can boost up the noise power significantly, which in turn results in performance degradation. To solve this, MMSE minimizes the mean squared-error, i.e. $J(\mathbf{W}) = E\{(x-\hat{x})*(x-\hat{x})\}$, with respect to $\mathbf{W}$ [13], [14].

4.2 Non- linear Detectors

4.2.1 VBLAST

A popular nonlinear combining approach is the vertical Bell labs layered space time algorithm (VBLAST) [9],[10] also called Ordered Successive Interference Cancellation (OSIC). It uses the detect-and-cancel strategy similar to that of decision-feedback equalizer. Either ZF or MMSE can be used for detecting the strongest signal component used for interference cancellation. The performance of this procedure is generally better than ZF and MMSE. VBLAST provides a suboptimal solution with lower computational complexity than ML. However, the performance of VBLAST is degraded due to error propagation.

4.2.2 ML Detector

Maximum Likelihood detector is optimal but computationally very expansive. ML detection is not practical in large MIMO systems.

5   ACO for Spatial Multiplexing System

ACO is an attractive technique that is very effective in solving optimization problems that have discrete and finite search space. Since the optimal MIMO detection problem involves a search process across the finite number of possible solutions, ACO is an ideal candidate to solve this problem.

5.1   Ant Colony Optimization (ACO)

ACO is based on the behavior of a colony of ants searching for food. In this approach, several artificial ants perform a sequence of operations iteratively. Within each iteration, several ants search in parallel for good solutions in the solution space. One or more ants are allowed to execute a move iteratively, leaving behind a pheromone trail for others to follow. An ant traces out a single path, probabilistically selecting only one element at a time, until an entire solution vector is obtained. In the following iterations, the traversal of ants is guided by the pheromone trails, i.e., the stronger the pheromone concentration along any path, the more likely an ant is to include that path in defining a solution. The quality of produced solution is estimated via a cost function in each iteration. This estimate of a solution quality is essential in determining whether or not to deposit pheromone on the traversed path.

As the search progresses, deposited pheromone dominates ants' selectivity, reducing the randomness of the algorithm. Therefore, ACO is an exploitive algorithm that seeks solutions using information gathered previously, and performs its search in the vicinity of good solutions. However, since the ant's movements are stochastic,

ACO is also an exploratory algorithm that samples a wide range of solutions in the solution space.

5.2 Binary Ant System (BAS)

1) Solution construction: In BAS, artificial ants construct solutions by traversing the mapping graph as shown in Fig.2 below.

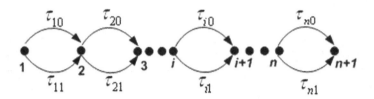

Fig. 2. Routing Diagram for Ants in BAS

A number of n_a ants cooperate together to search in the binary solution domain per iteration. Each ant constructs its solution by walking sequentially from node 1 to node n+1 on the routing graph shown above. At each node i, ant either selects upper path i_0 or the lower path i_1 to walk to the next node $i+1$. Selecting i_0 means $x_i=0$ and selecting i_1 means $x_i=1$. The selecting probability is dependent on the pheromone distribution on the paths:

$$p_{is} = \tau_{is}(t), i = 1,...,n, s \in \{0,1\} \tag{6}$$

here 't' is the number of iterations.

2) Pheromone Update: The algorithm sets all the pheromone values as $\tau_{is}(0) = 0.5$, initially but uses a following pheromone update rule:

$$\tau_{is}(t+1) \leftarrow (1-\rho)\tau_{is}(t) + \rho \sum_{x \in S_{upd}|is \in x} w_x \tag{7}$$

Where S_{upd} is the set of solutions to be intensified; w_x are explicit weights for each solution $x \in S_{upd}$, which satisfying $0 \leq w_x \leq 1$ and $\sum_{x \in Supd|is \in x} w_x = 1$. The evaporation parameter ρ is initially as ρ_0, but decreases as $\rho \leftarrow 0.9\rho$ every time the pheromone re-initialization is performed. S_{upd} consists of three components: the global best solution S^{gb}, the iteration best solution S^{ib}, and the restart best solution $S^{rb.}$ w_x combinations are implemented according to the convergence status of the algorithm which is monitored by convergence factor cf, given by:

$$cf = \sum_{i=1}^{n} |\tau_{i0} - \tau_{i1}| / n \tag{8}$$

The pheromone update strategy in different values of cf, are given in table-1, here w_{ib}, w_{rb} and w_{gb} are the weight parameters for S^{ib}, S^{rb} and S^{gb} respectively, cf_i, $i=1,...,5$ are threshold parameters in the range of [0,1]. When $cf > cf_5$, the pheromone re-initialization is preformed according to S^{gb}.

5.3 BA-MIMO Detection Algorithm

The major challenge in designing BAS based MIMO detector is the selection of effective fitness function which is problem dependent and perhaps is the only link between the real world problem and the optimization algorithm. The basic fitness function used by this optimization algorithm to converge to the optimal solution is (5). Choice of initial solution plays a vital role in the fast convergence of the optimization algorithm to a suitable solution. We therefore, make a start with the ZF or VBLAST input to the BA-MIMO detection algorithm. The proposed detection algorithm is described as follows:

1) Take the output of ZF or VBLAST as initial input to algorithm instead of keeping random values, such that $x_i \in \{0,1\}$. Number of nodes n visited by n_a ants is bxN_t i.e ML search space size (x_i). Here x_i represents the bit strings of the detected symbols at the receiver and $i = 1$ to n.

2) The probability of selecting $x_i = 0$ or 1 depends upon the pheromone deposited according to (7). Where $\tau_{is}(0) = 0.5$ for equal initial probability.

3) Evaluate the fitness of solution based on (5):

$$f = \left\| y - Hx \right\|^2 \tag{9}$$

Minimum Euclidean distance for each symbol represents the fitness of solution. Effect on the Euclidean distance due to x_i measured.

4) Pheromone update based on (7) is performed. S_{upd} that consists of S^{gb}, S^{ib}, and S^{rb} is calculated with weights w_x based on cf (8) and table-1.

5) Goto step-3 until maximum number of iterations is reached. The solution gets refined iteratively.

As $cf \rightarrow 0$, the algorithm gets into convergence, once $cf > cf_5$, the pheromone re-initialization procedure is done according to S^{gb}.

$$\tau_{is} = \tau_H \quad if \quad is \in S^{gb}$$
$$\tau_{is} = \tau_L \quad otherwise \tag{10}$$

where τ_H and τ_L are the two parameters satisfying $0 < \tau_L < \tau_H < 1$ and $\tau_L + \tau_H = 1$. The algorithm parameters are set as : $x_i = bN_t$, $\tau_0 = .5$, $\tau_H = .65$ and $\rho_0 = 0.3$.

Table 1. Pheromone Update Strategy for BA-MIMO system

	$cf < cf_1$	$cf < [cf_1, cf_2)$	$cf < [cf_2, cf_3)$	$cf < [cf_3, cf_4)$	$cf < [cf_4, cf_5)$
w_{ib}	1	2/3	1/3	0	0
w_{rb}	0	1/3	2/3	1	0
w_{gb}	0	0	0	0	1

6 Performance Analysis of BA-MIMO Detection Technique

6.1 BER Versus SNR Performance

For performance comparison, we present simulation results of the proposed method with existing detection techniques for the spatial multiplexing systems. 128 sub-carriers and cyclic prefix of length 32 are used. The SNR (E_b/N_o) is the average Signal to noise ratio per antenna (P/σ_v^2) where P is the average power per antenna and σ_v^2 is the noise variance. The simulation environment assumes Rayleigh fading channel with no correlation between sub-channels. The channel is assumed to be quasi-static for each symbol, but independent among different symbols. Perfect sampling and carrier frequency offset synchronization are assumed. An average of no less than 10,000 simulations is obtained for each result in order to report statistically relevant results.

A 3x3 (N_txN_r), 4x4, and 4-QAM 6x6 MIMO system is simulated. The symbols x_i and number of algorithm iterations (N_{itr}) depends upon N_t and QAM constellation size. For 3x3, 4-QAM system, x_i equals 6 and it grows to 12 for 6x6, 4-QAM system. N_{itr} is kept in the range of 10 to 20 in our simulations. Iterations are according to the system requirements. Larger N_{itr} can result in better BER at the cost of complexity. However, the algorithm reaches saturation after a certain number of iterations and therefore N_{itr} needs to be tuned carefully. Optimum N_{itr} value is taken after a number or trials to find the best BER with least complexity.

Fig. 3 present the BER versus E_b/N_o performance of proposed detector compared with ML and VBLAST detectors for 3x3 ($N_{itr}=10$). At 10^{-3} BER, the proposed BAV algorithm (with VBLAST as initial guess) and BAZ algorithm (with ZF as initial guess), result in 3-dB and 5-dB degraded performance in comparison with ML. However, in comparison with VBLAST, both the BAV and BAZ show 7-dB and 5-dB enhanced performance, respectively.

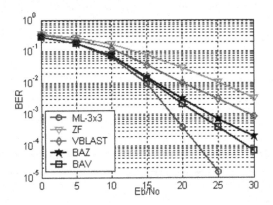

Fig. 3. BER versus SNR performance for a 3x3 system

In a 4x4 system in Fig. 4 ($N_{itr}=13$), the BER gain of both BAZ and BAV detectors, in comparison with VBLAST increase by 7-dB and 8-dB, respectively. The BER performance of BAZ and BAV detectors lower by 4-dB and 3-dB from ML but with some complexity reduction (discussed next).

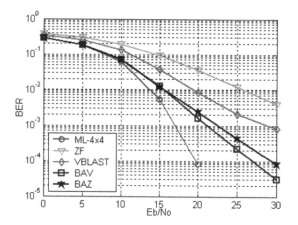

Fig. 4. BER versus SNR performance for a 4x4 system

For a 6x6 ($N_{itr}=18$) system in Fig. 5, the BER improvement in comparison to VBLAST for both the BAZ and BAV detectors is 7-dB and 8-dB, respectively. However, ML has 7-dB (approximately) superior BER performance but its complexity is also significant.

Increase in system configuration (N_txN_r), results in exponential increase of search space, therefore more algorithm iterations are required to converge to near-optimal solution. A trade off between systems BER performance and iterations has to be maintained according to the system requirement and priority.

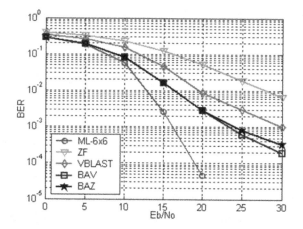

Fig. 5. BER versus SNR performance for a 6x6 system

6.2 Computational Complexity Comparison

Here we examine the computational complexity of the reported detector and compare it with ML and VBLAST detectors. As the hardware cost of each algorithm is implementation-specific, we try to provide a rough estimate of complexity in terms of number of complex multiplications. The computational complexity is computed in terms of the N_t, N_r and the constellation size M.

For ML detector as seen from (5) $M^{N_t}(N_r N_t)$ multiplications are required for matrix multiplication operation and additional $M^{N_t} N_r$ multiplications are needed for square operation. Therefore, ML complexity becomes:

$$\gamma_{ML} = N_r(N_t + 1)M^{N_t} \tag{11}$$

In case of ZF, the pseudo-inverse of matrix $(\mathbf{H}^H\mathbf{H})^{-1}\mathbf{H}^H$ takes $4N_t^3 + 2N_t^2 N_r$ multiplications [15]. Therefore ZF complexity becomes:

$$\gamma_{ZF} = 4N_t^3 + 2N_t^2 N_r \tag{12}$$

For VBLAST the pseudo-inverse matrix is calculated N_t times with decreasing dimension. In addition, the complexity of ordering and interference canceling is $\sum_{i=0}^{N_t-1}[N_t(N_t - i) + 2N_t]$. Therefore, total complexity of VBLAST (γ_{VBLAST}) results in.

$$\gamma_{VBLAST} = \sum_{i=0}^{N_t}(4i^3 + 2N_r i^2) + \sum_{i=0}^{N_t-1}[N_t(N_t - i) + 2N_t] \tag{13}$$

$$= N_t^4 + (5/2 + 2/3 N_r)N_t^3 + (7/2 + N_r)N_t^2 + 1/3 N_t N_r \tag{14}$$

For BA-MIMO detector, first fitness using (5) in x_i is calculated. Therefore, Multiplication complexity ($\gamma_{BA-MIMO}$) becomes,

$$\gamma_{BA-MIMO} = x_i(N_t N_r) \tag{15}$$

Pheromone update requires μ_p additional multiplications per iteration with from (7). Therefore μ_p becomes 2, the complexity becomes,

$$\gamma_{BA-MIMO} = x_i(N_t N_r + \mu_p) \tag{16}$$

This procedure is repeated N_{itr} (same as t) times to converge to the near-optimal BER performance. Therefore,

$$\gamma_{BA-MIMO} = x_i(N_t N_r + \mu_p)N_{itr} \tag{17}$$

The Proposed detector takes initial solution guess as ZF or VBLAST output therefore, it is added into get the resultant complexity of BA-MIMO detector.

From (5),(11) it is observed that the complexity of ML is exponential with N_t and M. ML complexity for a 4-QAM 4x4 system is 5120 and it grows to 4.7 M for 8x8 system. This increase is even significant with higher order modulation schemes.

Computational complexity of VBLAST for 4-QAM 4x4 and 6x6 systems computed from (14) is 712 and 3054, respectively. The complexity of proposed detector with ZF initialization for 4x4 and 6x6 configurations is 2256 and 9504. However, the complexity with VBLAST input comes out to be 2584 and 11262, respectively. This complexity estimate is only meaningful in the order of magnitude sense since it is based on the number of complex multiplications. The above complexity is estimated on subcarrier-by-subcarrier for MIMO-OFDM system.

6.3 Performance-Complexity Trade-Off

A reasonable performance-complexity trade-off exists when a comparison of the proposed detector is drawn with ML and VBLAST detectors. Table 2 draws a comparison of the proposed detector with optimal ML and VBLAST detectors. Compared to ML, complexity reduction of the proposed detector given by (γ_{ML} - $\gamma_{BA\text{-}MIMO}$)/γ_{ML} is significant. In comparison to VBLAST, for 4x4 system, the proposed detector with ZF input results in 7-dB increased BER performance at the cost of increased 68 % computational time. However, as compared to ML, the complexity is still 56 % lower with 4-dB lesser BER performance. This computational complexity saving increases to as high as 94% at the cost of 7-dB lesser BER performance for a 6x6 system with ZF input.

Table 2. 4-QAM $N_t x N_r$ systems - performance complexity trade-off

Performance comparison between ML, VBLAST and proposed detectors	**4x4** (n_a=8, N_{itr}=13, μ_p=2)		**6x6** (n_a=12, N_{itr}=18, μ_p=2)	
	Performance (at 10^{-3} BER)	Computational Complexity	Performance (at 10^{-3} BER)	Computational Complexity
BAZ and ML	4-dB less	56% less	7-dB less	94% less
BAV and ML	3-dB less	50% less	7-dB less	93% less
BAV and VBLAST	8-dB more	71% more	8-dB more	80% more
BAZ and VBLAST	7-dB more	68% more	7-dB more	67% more

7 Conclusions

In this paper, a binary ant system assisted symbol detection in a spatial multiplexing system is reported. The algorithms simple model, lesser implementation complexity, resistance to being trapped in local minima, convergence to reasonable solution in fewer iterations and exploratory-exploitive search approach makes it a suitable candidate for real-time wireless communications systems. This algorithm shows promising results when compared with the optimal ML and traditional VBLAST detectors. This BA-MIMO symbol detection mechanism approaches near-optimal performance with much reduced computational complexity, especially for complex systems with multiple transmitting antennas, where conventional ML detector is computationally expensive and impractical to implement. When compared to

VBLAST detector the proposed unconventional detection method results in enhanced BER performance but at the cost of increase in complexity. The simulation results suggest that the proposed detector in a 6x6 spatial multiplexing system improves VBLAST BER performance by 7-dB. However, the ML complexity is reduced by 94% with a reasonable BER performance.

References

[1] Dorigo, M., Maniezzo, V., Colorni, A.: Positive feedback as a search strategy. Technical report, Dipartimento di Elettronica e Informatica, Politecnico di Milano, IT (1991)

[2] Dorigo, M., Gambardella, L., Middendorf, M., Stutzle, T.: Guest editorial: special section on ant colony optimization (2002)

[3] Dorigo, M., Gambardella, L.: Ant colony system: a cooperative learning approach to the traveling salesman problem. IEEE Transactions on Evolutionary Computation 1, 53–66 (1997)

[4] Kong, M., Tian, P.: Introducing a Binary Ant colony Optimization. In: Fifth International Workshop on Ant Colony Optimization and Swarm Intelligence (2006)

[5] Khan, A.A., Naeem, M., Shah, S.I.: A Particle Swarm Algorithm for Symbols Detection in Wideband Spatial Multiplexing Systems. In: GECCO-07 accepted for publication in Genetic and Evolutionary Computation Conference (2007)

[6] Goldsmith, A., Jafar, S.A., Jindal, N., Vishwanath, S.: Capacity limits of MIMO channels. IEEE Journal on Selected Areas in Communications 20(5) (2003)

[7] Mujtaba, S.: TGn Sync Proposal Technical Specification. Agere Systems Inc. (2005), http://www.tgnsync.org/techdocs/11-04-0889-06-000n-tgnsyncproposal-technical-specification.doc

[8] Bolcskei, H., Gesbert, D., Papadias.: Space-Time Wireless Systems: From Array Processing to MIMO Communications. Cambridge University Press, Cambridge (2005)

[9] Foschini, G.J.: Layered space-time architecture for wireless communication in a fading environment when using multiple antennas. Bell Labs Technical Journal 1, 41–59 (1996)

[10] Foschini, G.J., Gans, M.J.: On limits of wireless communications in a fading environment when using multiple antennas. Wireless Personal Communications 6(3), 311–335 (1998)

[11] Choi, W.J., Negi, R., Cioffi, J.M.: Combined ML and DFE decoding for the V-BLAST system. In: Proc. IEEE International Conference on Communications 2000, pp. 1243–1248. IEEE Computer Society Press, Los Alamitos (2000)

[12] Choi, W.J., Cheong, K.W., Cioffi, J.M.: Iterative soft interference cancellation for multiple antenna systems. In: Proc.IEEE Wireless Communications and Networking Conference 2000, pp. 304–309. IEEE Computer Society Press, Los Alamitos (2000)

[13] Haykin, S.: Adaptive Filter Theory, 3rd edn. Prentice-Hall, Englewood Cliffs (1996)

[14] Sayed, A.: Fundamentals of Adaptive Filtering, Wiley-IEEE Press (2003)

[15] Golub, G.H., Loan, C.F.V.: Matrix Computations, 3rd edn. John Hopkins University Press, Baltimore, MD (1996)

Quantum Authenticated Key Distribution*

Naya Nagy and Selim G. Akl

School of Computing, Queen's University
Kingston, Ontario
Canada K7L 3N6
{nagy,akl}@cs.queensu.ca

Abstract. Quantum key distribution algorithms use a quantum communication channel with quantum information and a classical communication channel for binary information. The classical channel, in all algorithms to date, was required to be authenticated. Moreover, Lomonaco [8] claimed that authentication is not possible using only quantum means. This paper reverses this claim. We design an algorithm for quantum key distribution that does authentication by quantum means only. Although a classical channel is still used, there is no need for the channel to be authenticated. The algorithm relies on two protected public keys to authenticate the communication partner.

Keywords: quantum key distribution, authentication, entanglement.

1 Introduction

Most cryptosystems commercially used today, rely on the principles of public key cryptography. Invariably, such cryptosystems aim to offer the means of exchanging secret messages securely and reliably. Suppose two entities, Alice and Bob, want to exchange secret messages; specifically, Bob prepares a secret message to be sent to Alice. Unfortunately, all they have available is a classical insecure communication channel. This means, a malevolent third party, Eve, makes every effort to ruin the secrecy or content of Bob's message. Eve can listen to the communication channel to find out the content of Bob's message. And also, Eve can tamper with Bob's message, adding, deleting or editing parts of the message. The security of the public key cryptosystem relies on the difficulty of inverting particular algebraic functions, also called "one-way" functions.

Secure communication is achieved using two types of keys: a public key and a private key. If Bob wants to send a secret message to Alice, he uses the public key to encrypt the message. Alice then reads the message after using her private key for decoding. There are a few characteristics worth mentioning about the two keys implied in this communication. Alice's private key is secret, and not shared with anybody else. In particular, Bob does not need to know Alice's private key. This is a major advantage, as the private key is never seen on any

---

* This research was supported by the Natural Sciences and Engineering Research Council of Canada.

S.G. Akl et al.(Eds.): UC 2007, LNCS 4618, pp. 127–136, 2007.
© Springer-Verlag Berlin Heidelberg 2007

communication channel and therefore, its secrecy is ensured. The public key is available to anybody. Bob needs to know it, and also an eavesdropper, Eve, has access to it. In order for the protocol to work, the public key is guaranteed to be protected. This means, there is a consensus about the public key value. Both Bob and Alice are sure that they use the correct, same public key. Eve cannot masquerade as Alice and change the value of the public key, making Bob use a false public key to encrypt his message. This feature of the public key is important. Our authentication protocol for quantum key distribution makes use of this property of the protected public key. It is crucial in both the classical sense of authentication protocols, as well as in our protocol, that such a public key can be published with the guarantee that the key *is and remains* protected from masquerading.

The security of the public key distribution protocol relies on the theoretically unproven assumption that factoring large numbers is intractable on classical computers. As described in [7], quantum computers can break some of the best public key cryptosystems.

Quantum cryptography aims to design mechanisms for secret communication with higher security than protocols based on the public key approach. Privacy of a message and its credibility is well satisfied in a private key cryptosystem setting. Alice and Bob share one and the same secret key, k_s. Bob uses the secret key for encryption and Alice consequently decrypts the message with the same key. As long as k_s is unknown to anybody else, the secrecy of the communication is satisfied. There exist various encryption / decryption functions using k_s, such that the encrypted message reveals no information whatsoever about the content of the message, provided the key k_s is unavailable.

Quantum key distribution protocols establish secret keys via insecure quantum and/or classical channels. Existing quantum key distribution algorithms generally use two communication channels between Alice and Bob: a quantum channel which transmits qubits and a classical channel for classical binary information. The classical channel is used to communicate measurement strategy, or the basis for measurement, and to check for eavesdropping.

Quantum key distribution protocols may derive their efficiency from different quantum properties. The first protocol developed by Charles Bennett and Gilles Brassard, known as the BB84 protocol [2], relies on measuring qubits in two different orthonormal bases. The same idea applies to any two nonorthogonal bases [1]. In [5] the quantum key distribution algorithm is derived from the quantum Fourier transform. Based on the property of entanglement, Artur Ekert [4] gave a quantum key distribution solution using entangled qubits to be shared by Alice and Bob. A simpler version with qubits entangled in the same way, namely in the Bell states, is described in [3].

Note that all quantum key distribution algorithms mentioned above require that the classical channel be authenticated. Authentication is supposed to be done by classical means. The authenticated classical channel prevents Eve from masquerading as someone else and tamper with the communication. It was claimed by Lomonaco [8] that authentication is not possible in quantum computation, that for any secure quantum communication a classical authentication scheme needs to be used.

As will be clear from the algorithm described in this paper, authentication of a quantum communication protocol can be done by the quantum protocol itself. The classical channel in our algorithm is not authenticated. Yet Alice and Bob do have an authentication step at the end of the protocol, with the help of protected public keys. Authentication is derived from the quantum algorithm itself and can catch any masquerading over the classical channel.

Shi et. al [9] describe in their paper a quantum key distribution algorithm that does not use a classical channel at all. Authentication is done by a trusted authority, that provides the entangled qubits to Alice and Bob. In our paper, such a trusted authority is not needed. The entangled qubits may come from an insecure source.

The rest of the paper is organized as follows: Section 2 defines entanglement and describes the particular entanglement based on phase incompatibility used by our algorithm. Section 3 describes the algorithm with authentication and security checking. Section 4 concludes the paper and offers some future directions for investigation.

2   Entangled Qubits

The key distribution algorithm we present in the following sections relies on entangled qubits. Alice and Bob, each possess one of a pair of entangled qubits. If one party, say Alice, measures her qubit, Bob's qubit will collapse to the state compatible with Alice's measurement.

The algorithms mentioned above [4,3,9], all rely on Bell entangled qubits. The qubit pair is in one of the four Bell states:

$$\frac{1}{\sqrt{2}}(|00\rangle \pm |11\rangle)$$

$$\frac{1}{\sqrt{2}}(|01\rangle \pm |10\rangle)$$

Suppose Alice and Bob share a pair of entangled qubits described by the first Bell state:

$$\frac{1}{\sqrt{2}}(|00\rangle + |11\rangle)$$

Alice has the first qubit and Bob has the second. If Alice measures her qubit and sees a 0, then Bob's qubit has collapsed to $|0\rangle$ as well. Bob will measure a 0 with certainty, that is, with probability 1. Again, if Alice measures a 1, Bob will measure a 1 as well, with probability 1. The same scenario happens if Bob is the first to measure his qubit.

Note that any measurement on one qubit of this entanglement collapses the other qubit to a *classical* state. This property is specific to all four Bell states and is then exploited by the key distribution algorithms mentioned above: If Alice measures her qubit, she *knows* what value Bob will measure. The entanglement employed in this paper and algorithm does not have this property directly.

2.1 Entanglement Caused by Phase Incompatibility

Let us look now at an unusual form of entanglement. Consider the following ensemble of two qubits:

$$\phi = \frac{1}{2}(-|00\rangle + |01\rangle + |10\rangle + |11\rangle)$$

The ensemble has all four components, $|00\rangle$, $|01\rangle$, $|10\rangle$, and $|11\rangle$, in its expression. And yet, this ensemble is entangled.

Consider the following proof. Suppose the ensemble ϕ is not entangled. This means ϕ can be written as a scalar product of two independent qubits:

$$\phi = \frac{1}{2}(\alpha_1|0\rangle + \beta_1|1\rangle)(\alpha_2|0\rangle + \beta_2|1\rangle)$$

Matching the coefficients from each base vector, we have the following conditions:

1. $\alpha_1\alpha_2 = -1$
2. $\alpha_1\beta_2 = 1$
3. $\alpha_2\beta_1 = 1$
4. $\beta_1\beta_2 = 1$

The multiplication of conditions 1 and 4 have the result: $\alpha_1\alpha_2\beta_1\beta_2 = -1$. From conditions 2 and 3, we have: $\alpha_1\alpha_2\beta_1\beta_2 = 1$. This is a contradiction. The product $\alpha_1\alpha_2\beta_1\beta_2$ cannot have two values, both $+1$ and -1. It follows that ϕ cannot be decomposed and thus the two qubits are entangled.

The entanglement of the ensemble is caused by the *signs* in front of the four base vector components. Thus, it is not that some vector is missing in the expression of the ensemble, but the phases of the base vectors keep the two qubits entangled.

2.2 Measurement

Let us investigate what happens to the ensemble ϕ, when the entanglement is disrupted through measurement.

If the first qubit q_1 is measured and yields $q_1 = |0\rangle = 0$ then the second qubit collapses to $q_2 = \frac{1}{\sqrt{2}}(-|0\rangle + |1\rangle)$. This is not a classical state, but a simple Hadamard gate transforms q_2 into a classical state. The Hadamard gate is defined by the matrix

$$H = \frac{1}{\sqrt{2}}\begin{bmatrix} 1 & 1 \\ 1 & -1 \end{bmatrix}$$

Applying the Hadamard gate to an arbitrary qubit, we have $H(\alpha|0\rangle + \beta|1\rangle) = \alpha\frac{|0\rangle+|1\rangle}{\sqrt{2}} + \beta\frac{|0\rangle-|1\rangle}{\sqrt{2}}$. For our collapsed q_2, we have $H(q_2) = H(\frac{1}{\sqrt{2}}(-|0\rangle + |1\rangle)) = -|1\rangle$. This is a classical 1.

The converse happens when qubit q_1 yields 1 through measurement. In this case q_2 collapses to $q_2 = \frac{1}{\sqrt{2}}(|0\rangle + |1\rangle)$. Applying the Hadamard gate transforms q_2 to $H(q_2) = H(\frac{1}{\sqrt{2}}(|0\rangle + |1\rangle)) = |0\rangle = 0$. Again this is a classical state 0.

It follows that by using the Hadamard gate, there is a clear correlation between the measured values of the first and second qubit. In particular, they always have opposite values.

A similar scenario can be developed, when the second qubit q_2 is measured first. In this case, the first qubit q_1, transformed by a Hadamard gate, yields the opposite value of q_2.

Note that the measurement of the qubits is assymmetric. Alice and Bob do not perform the same operation. This feature will be exploited when checking for eavesdropping.

3   The Algorithm

Alice and Bob wish to establish a secret key, to be used henceforth to encrypt / decrypt messages. One session is required to establish a binary secret key, called *secret*, such that Alice and Bob are in consensus about the value of the secret key. The secret key *secret* consists of n bits, $secret = b_1b_2...b_n$. Technically, to perform the algorithm, Alice and Bob need a classical communication channel, an array of entangled qubit pairs, and two protected public keys.

On the classical channel, classical binary information can be exchanged. The channel is *unprotected* and *not authenticated*. The channel, being unprotected, is sensitive to attacks of eavesdropping: Eve may attempt and successfully read information from the channel. Also, the channel, not being authenticated, is sensitive to masquerading: Eve may disconnect the channel and then talk to Alice pretending she is Bob, and talk to Bob pretending she is Alice.

The array of the entangled qubits has length l, it consists of l qubit pairs denoted $(q_{1A}, q_{1B}), (q_{2A}, q_{2B}), ..., (q_{lA}, q_{lB})$. The array is split between Alice and Bob. Alice receives the first qubit of each entangled qubit pair, namely q_{1A}, q_{2A}, ..., q_{lA}, and Bob receives the second half of the qubit pairs, q_{1B}, q_{2B}, ..., q_{lB}. The entanglement of the qubit pair is of the type described in the previous section, namely, phase incompatibility. The array of qubits is unprotected either. There is no guarantee that the qubits of a pair are indeed entangled, Eve may have disrupted the entanglement. Also, Eve may have masqueraded as either Alice or Bob, modifying the entangled qubits, such that Alice's qubit is actually entangled with a qubit in Eve's possession rather than Bob's, and the same holds for Bob.

Two public keys are needed by the algorithm. Alice has a public key key_A and Bob has a public key key_B. The two public keys key_A and key_B are independent. These keys are necessary for authentication. They have some characteristics that are different from the classical public keys. The keys are established *during* the computation. They are not known prior to the key distribution algorithm and are defined in value during the computation according to the measured values of some of the qubits. This means that the keys are available *after* the key distribution protocol. Consequently, the keys have to be posted after the algorithm, which is unlike the classical case, where a public key is known in advance. Also, the two public keys key_A and key_B are valid for one session, for one application of the key distribution algorithm. If Alice and Bob want to distribute a second secret key using the same algorithm, they will have to create new public keys, which are different in value from the public keys of the previous session.

The key distribution algorithm, like all quantum key distribution algorithms, develops the value of the secret key during the computation. Implicitly, the values of the public keys as well are developed *during* the computation. There exists no knowledge whatsoever about the values of the keys (secret and public) prior to running the algorithm.

The algorithm follows the steps below:

– **Step 1 - Establish the value of the secret key**

For each entangled qubit pair (q_{iA}, q_{iB}) in the array, the following actions are taken. On the classical channel, Alice and Bob decide randomly who is going to perform the first measurement.

Suppose it is Alice. Therefore, Alice measures her qubit q_{iA} thereby collapsing Bob's qubit q_{iB} to the state consistent with Alice's measurement. If Alice has measured a 0, $q_{iA} = 0$ then Bob's qubit has collapsed to $q_{iB} = \frac{1}{\sqrt{2}}(-|0\rangle + |1\rangle)$. If Alice's measurement resulted in $q_{iA} = 1$ then Bob's qubit collapsed to $q_{iB} = \frac{1}{\sqrt{2}}(|0\rangle + |1\rangle)$. Now, Bob transforms his qubit via a Hadamard gate. For Alice's $q_{iA} = 0$, Bob has a $Hq_{iB} = 1$, and conversely for Alice's $q_{iA} = 1$, Bob has a $Hq_{iB} = 0$. Bob now measures and his value will consistently be the complement of Alice's.

If Bob is the one who measures first his qubit q_{iB}, the procedure is simply mirrored. Alice now has to apply a Hadamard gate on her qubit, thus obtaining Hq_{iA}. Again Alice and Bob will have measured complementary binary digits.

Ideally, with no interference from Eve, be it through eavesdropping or masquerading, applying this measurement and Hadamard measurement on each qubit is enough to establish the secret key. After going through all the qubit pairs, Alice and Bob will have complements of the same binary number. This, for example Alice's binary number, is the established secret key.

– **Step 2 - Authentication and Eavesdropping Checking**

Some $2m$ qubits will be sacrificed for security checking, where $2m < l$. The secret key will be formed by the remaining $n = l - 2m$ qubits. Alice and Bob decide via the classical channel, the set of qubit indices to be sacrificed. Alice looks at the first m qubits, and forms a binary number with the values she reads. This is Alice's *public key*. Alice now publishes her public key, which key can be seen by Bob. As the public key is protected, Bob is certain that the public key is safe from masquerading. Note that this is the only step, where Bob is certain to have contact with Alice with no masquerading. Bob now compares Alice's public key with his own measured qubits. If these two binary numbers are complementary, then Bob concludes that he has been talking with Alice all the while in the previous step, and also that no eavesdropper has changed some of the qubit values. The same procedure applies to Bob's *public key* formed by the second m sacrificed qubits. If Bob or Alice have encountered a mismatch in the values checked, they discard the secret key and try again.

Let us analyze what Eve can do to get some knowledge about the secret key without being caught.

Note that Eve can have access to the qubits before they reach Alice and Bob, prior to the actual protocol. For example, Alice generates entangled qubit pairs. From each pair, she sends one qubit to Bob. The qubits are sent to Bob insecurely: this is the moment where Eve can act upon the qubits through measuring or exchanging qubits with her own. The initial distribution of entangled qubits is therefore insecure. Eve is allowed any intervention of her choice on the qubits. Alice and Bob have no guarantee that the qubits in their possession are indeed entangled with each other as expected. *But,* and this is essential for the security of the algorithm, once Alice and Bob, respectively, have their qubits in their possession, Eve has no more access to them. This is a perfectly reasonable assumption, that Alice and Bob can guard the qubits in their possession from outside attacks. Eve cannot measure or exchange a qubit that has already reached Alice and is in her possession. As such, *if* Alice and Bob have received a correctly entangled qubit pair before they have started the protocol, Eve has no chance to disrupt the entanglement afterwards. This means that Eve's attack on the entangled qubits can happen only prior to any decision concerning measurement.[1] Now, the advantage of an asymmetric measurement scheme can be clearly seen. Eve has no gain in measuring a qubit of the entanglement before the protocol is started. For example, if she does measure Alice's qubit directly, the later random measurement decision during the protocol may result in a Hadamard gate measurement for Alice, thus exposing Eve's intervention.

To evaluate a specific attack, let us consider that Alice's and Bob's qubits are not really entangled, but Eve has sent qubits of her own choice to both of them. Eve also can listen to the classical channel. The best she can do is send a classical 0 to Alice and a *Hadamard* 1 to Bob. Actually, all other combinations are equivalent to, or less advantageous than, this one. Alice and Bob decide who is to measure first once they already have the qubits. With 1/2 probability, Alice is measuring, in which case the readings are consistent. Bob measures first, again with probability 1/2. Bob will measure a 1 or 0 with equal probability. Then Alice transforms the classical 0 with a Hadamard gate, and also reads a

---

[1] This property is even stronger in our improved algorithm with no classical communication presented in the Conclusion section. Here, the decision on how to measure a qubit (directly or with a Hadamard gate) is not made on a classical communication channel, but in private by Alice and Bob independently.

0 or 1 with equal probability. This means that if Bob measures first, Alice and Bob will read the same value with 1/2 probability. As such, Eve is caught with 1/4 probability. If the number of qubits to be tested is large, this probability can be made arbitrarily large.

Note that, the classical channel does not need to be authenticated. Eve can masquerade, such that she completely severs all connection between Alice and Bob. In this case, Eve will establish one secret key with Alice, and another secret key with Bob. Unfortunately for Eve, she has no control over the value of the secret keys, as their values are determined probabilistically, through quantum measurement. Therefore, the two keys will necessarily differ. As Alice and Bob publish part of their secret keys as protected public keys, they will notice the difference and consequently discard the keys.

This algorithm features a few notable characteristics. Checking for eavesdropping and authentication happens in one and the same step. Until this checking phase, there is no certainty whatsoever about either the validity of the key, the validity of the classical connection or the quantum qubit entanglement. But then, the key is not *useful* or *used* before the checking step.

Another interesting feature is the way in which the two public keys are used in our case. In classical settings, the public key is established and known by both parties, before the algorithm or communication begins. By contrast, in this quantum distribution algorithm, the public keys are determined *during* the computation and they are available only after the secret key is established. The publishing of the public keys is the very last step of Alice and Bob's communication, whereas in previous algorithms this is the first step. The usage of the public keys is in reverse order by comparison with classical secret communications, such as those in the public key cryptosystems.

4 Conclusion

We have shown in this paper that quantum authentication in quantum key distribution can be done through the quantum protocol, and does not need to rely in any way on classical authentication procedures. Moreover, in our algorithm authentication is easily done with the help of two protected public keys.

The algorithm presented performs quantum key distribution based on entangled qubit pairs. The entanglement type is not of the generally used Bell states, but an unusual entanglement based on phase incompatibility.

The algorithm uses a quantum channel and a classical channel, but unlike all previous quantum key distribution algorithms, the classical channel is not authenticated. Authentication and security checking are done at the same time, *after* the algorithm, with the help of two public keys.

Specific to quantum key distribution algorithms, is the fact that the value of the secret key is not known prior to performing the distribution. The key is developed *during* the execution. Likewise, in our algorithm, we have the same behavior of the public keys. They are not known prior to the execution of the algorithm and are developed during the execution. The consequence is that the public keys are session specific, rather than permanent for one person. The public keys are distinct for each application of the quantum key distribution algorithm.

The algorithm presented here can be improved to work without a classical communication channel at all, as described in [6]. Alice and Bob communicate via the quantum channel of entangled qubits only. They each decide randomly, whether to measure directly or apply a Hadamard gate. Some, actually half, of the qubits will be measured correctly, with exactly one application of the Hadamard gate. The other half will be measured incorrectly, either with no Hadamard application or with two. The incorrectly measured qubits will be discarded. The very promising feature of the measurement policy here is that there exists no knowledge about how the qubits will be measured before they are actually measured. Alice and Bob decide randomly, on the spot, how to measure the qubit they are about to measure. There is no prior rule, that Eve might get knowledge of. Because the measurement is asymmetric, Eve cannot predict which measurement scheme will be followed by Alice and Bob. The tradeoff of this random measuring is a doubling in the number of quantum bits used for a final secret key of the same length. Also, the meaning of the public keys is more complex, as it incorporates information about the random measurements.

As shown in this paper, quantum authentication can be done with the help of a variation of protected public keys. This might not be the only solution. It is an open problem what other structures can support authentication of quantum channels.

The principle of checking and authenticating at the end of the protocol with public keys, is not restricted to the algorithm described here. The same type of public keys, namely per session keys, posted after the execution of the main body of the algorithm, can be successfully used in authenticating other types of algorithms. This is also a direction worth investigating.

Acknowledgments. The authors greatly appreciate the contribution of their colleague Marius Nagy in drawing their attention to entanglement based on phase incompatibility.

References

1. Bennett, C.H.: Quantum cryptography using any two nonorthogonal states. Physical Review Letters 68(21), 3121–3124 (1992)
2. Bennett, C.H., Brassard, G.: Quantum cryptography: Public key distribution and coin tossing. In: Proceedings of IEEE International Conference on Computers, Systems and Signal Processing, Bangalore, India, pp. 175–179. IEEE, New York (December 1984)
3. Bennett, C.H., Brassard, G., Mermin, D.N.: Quantum cryptography without Bell's theorem. Physical Review Letters 68(5), 557–559 (1992)
4. Ekert, A.: Quantum cryptography based on Bell's theorem. Physical Review Letters 67, 661–663 (1991)
5. Nagy, M., Akl, S.G.: Quantum key distribution revisited. Technical Report 2006-516, School of Computing, Queen's University, Kingston, Ontario (June 2006)

6. Nagy, N., Akl, S.G.: Authenticated quantum key distribution without classical communication. In: Workshop on Unconventional Computational Problems, Sixth International Conference on Unconventional Computation, Kingston, Ontario (August 2007)
7. Nielsen, M.A., Chuang, I.L.: Quantum Computation and Quantum Information. Cambridge University Press, Cambridge, UK (2000)
8. Lomonaco, Jr., S.J.: A Talk on Quantum Cryptography or How Alice Outwits Eve. In: Proceedings of Symposia in Applied Mathematics, vol. 58, pp. 237–264. Washington, DC (January 2002)
9. Shi, B.-S., Li, J., Liu, J.-M., Fan, X.-F., Guo, G.-C.: Quantum key distribution and quantum authentication based on entangled states. Physics Letters A 281(2-3), 83–87 (2001)

The Abstract Immune System Algorithm

José Pacheco[1,*] and José Félix Costa[2,3]

[1] Departamento de Informática, Faculdade de Ciências e Tecnologia
Universidade Nova de Lisboa,
Quinta da Torre, 2829-516 Caparica, Portugal
`jddp@di.fct.unl.pt`
[2] Department of Mathematics, Instituto Superior Técnico
Universidade Técnica de Lisboa
Lisboa, Portugal
`fgc@math.ist.utl.pt`
[3] Centro de Matemática e Aplicações Fundamentais do Complexo Interdisciplinar
Universidade de Lisboa
Lisbon, Portugal

Abstract. In this paper we present an *Abstract Immune System Algorithm*, based on the model introduced by Farmer *et al*, inspired on the theory of Clonal Selection and Idiotypic Network due to Niels Jerne. The proposed algorithm can be used in order to solve problems much in the way that Evolutionary Algorithms or Artificial Neural Networks do. Besides presenting the Algorithm itself, we briefly discuss its various parameters, how to encode input data and how to extract the output data from its outcome. The reader can do their own experiments using the workbench found in the address `http://ctp.di.fct.unl.pt/~jddp/immune/`.

1 Introduction

Biological studies have always constituted a large pool of inspiration for the design of systems. In the last decades, two biological systems have provided a remarkable source of inspiration for the development of new types of algorithms: neural networks and evolutionary algorithms.

In recent years, another biological inspired system has attracted the attention of researchers, the immune system and its powerful information processing capabilities (e.g., [21,10]). In particular, it performs many complex computations in a highly parallel and distributed fashion. The key features of the immune system are: pattern recognition, feature extraction, diversity, learning, memory, self-regulation, distributed detection, probabilistic detection, adaptability, specificity, etc. The mechanisms of the immune system are remarkably complex and poorly understood, even by immunologists. Several theories and mathematical models have been proposed to explain the immunological phenomena [19,16]. There is also a growing number of computer models to simulate various components of the immune system and the overall behaviour from the biological point

---

* Corresponding author.

S.G. Akl et al.(Eds.): UC 2007, LNCS 4618, pp. 137–149, 2007.
© Springer-Verlag Berlin Heidelberg 2007

of view [7,9]. Those approaches include differential equation models, stochastic differential equation models, cellular-automata models, shape-space models, etc.

The models based on the immune system principles, such as the Clonal Selection Theory [6,12], the immune network model [15,8], or the negative selection algorithm [14], have been finding increasing applications in science and engineering [9]: computer security, virus detection, process monitoring, fault diagnosis, pattern recognition, etc.

Although the number of specific applications confirms the interest and the capabilities of this principles, the lack of a general purpose algorithm for solving problems based on them contrasts with the major achievements with other Biologically inspired models. In this paper we present a modest *Abstract Immune System Algorithm*. Our paper is focused on describing the algorithm itself and how to use it.

2   The Biological Model

The immune system is a very complex system with several functional components [18,13]. It is constituted by a network of interacting cells and molecules which recognize foreign substances. These foreign substances are called *antigens*. The molecules of the immune system that recognize antigens are called *antibodies*. An antibody does not recognize an antigen as a whole object. Instead, it recognizes small regions called *epitopes*. An antibody recognizes an antigen if it binds to one of its epitopes. The binding region of an antibody is called the *paratope*. The strength and the specificity of the interaction between antibody and antigen is measured by the affinity of the interaction. The affinity depends on the degree of complementarity in shape between the interacting regions of the antibody and the antigen. A given antibody can typically recognize a range of different epitopes, and a given epitope can be recognized by different antibody types. Not only do antibodies recognize antigens but they also recognize other antibodies if they have the right epitope. An epitope characteristic for a given antibody type is called an *idiotope*. Antibodies are produced by cells called *B-lymphocytes*. B-lymphocytes differ in the antibodies that they produce. Each type of antibody is produced by a corresponding lymphocyte which produces only this type of antibody. When an antibody on the surface of a lymphocyte binds another molecule (antigen or other antibody), the lymphocyte is stimulated to clone and to secrete free antibodies. In contrast, lymphocytes that are not stimulated die after days. Thus, a selection process is at work here where those antibodies that are stimulated by antigens or antibodies are amplified, while the other antibodies die out.

There exist several theories to explain the dynamics of the immune system. One of the most popular theories is the *immune network model* proposed by Niels Jerne [15]. Jerne hypothesized that the immune system is a regulated network of molecules and cells that recognize one another even in the absence of antigen. Such networks are often called *idiotypic networks* which present a mathematical framework to illustrate the behaviour of the immune system. His theory is modeled with a system of differential equations which simulates the dynamics of lymphocytes, the increase or decrease of the concentration of lymphocyte clones and the corresponding immunoglobins. The idiotypic network hypothesis is based on the concept that lymphocytes are not isolated, but communicate with each other through interaction among antibodies. Jerne suggested that during an immune response antigens will induce the creation of a first set of antibodies. These antibodies would then act as antigens and induce a second set of *anti-idiotypic* antibodies. The repetition of this process produces a network of lymphocytes that recognize one another. With this hypothesis Jerne explains the display of memory by the immune system.

Other important issues are the incorporation of new types in the immune network and the removal of old types. The autoregulation of the immune system keeps the total number of different types roughly constant. Although new types are created continously, the probability that a newly created type is incorporated in the immune system is different for different types. In natural evolution, the creation of new individual result from the mutation and crossover of the individuals in the population. In the immune system, new antibody types are created preferably in the neighbourhood of the existing high-fit antibody types. The biased incorporation into the immune network of randomly created species is called the *recruitment strategy* [5,3,4].

Also fundamental for the present formulation is clonal selection theory [6,12]. This theory states that there is a bisimilar relation between lymphocytes and receptors, hence each kind of lymphocyte has only one kind of receptor. To explain the predominance or the disappearing of certain kinds of lymphocytes, the theory assumes a highly diverse initial population, composed by randomly distributed lymphocytes. The recognition of antigens produces a reaction of lymphocytes are stimulating their reproduction, through the production of clones. The measure of the adaptability of lymphocytes, is proportional to the intensity of the reaction, giving more chances of survival to those kinds of lymphocytes more stimulated on a given environment. This natural selection process together with the high rate of mutation found on the reproduction stage, induces an increase of the concentration of the lymphocytes with the highest molecular affinity with the form they try to eliminate.

3   The Algorithm

The algorithm presented here follows closely the model proposed by Farmer *et al* [11] for the dynamics of a system based on the idiotypic network model. The dynamics of the model is described using a set of differential equations of the form:

$$\dot{x}_i = c \left[\sum_{j=1}^{N} m_{ji} x_j x_i - k_1 \sum_{j=1}^{N} m_{ij} x_i x_j + \sum_{j=1}^{n} m_{ji}^* y_j x_i \right] - k_2 x_i$$

where N is the number of antibody types, with concentrations, respectively, x_1, x_2, ..., x_N, and n is the number of antigen types with concentrations y_1, y_2, ...,

y_n; m_{ji}^* correspond to the affinity between the paratope of the antibody i and the epitope of the antigen j; m_{ji} correspond to the affinity between the paratope of the antibody i and the epitope of the antibody j.

The first term represents the stimulation of the paratope of an antibody of type i by the epitope of an antibody of type j. The second term represents the suppression of the antibody of type i when its epitope is recognized by the paratope of type j. The constant k_1 represents the possible inequality between stimulation and suppression. The third term models the stimulation provided by the recognition of the antigen j (with concentration y_j) by the antibody of type i (with concentration x_i). The final term models the tendency of cells to die in the absence of any interaction, at a rate determined by k_2. The parameter c is a rate constant that depends on the number of collisions per unit of time and the rate of antibody production stimulated by a collision.

Since we want to model the evolution of a population of potential solutions of a problem, we consider the following set of equations:

$$\dot{x}_i = c \left[\sum_{j=1}^{N} m_{ji} x_j x_i - k_1 \sum_{j=1}^{N} m_{ij} x_i x_j \right] - k_2 x_i,$$

ignoring the antigens.

With some further simplification we get the equation:

$$\dot{x}_i = \left[c \sum_{j=1}^{N} (m_{ji} - k_1 m_{ij}) x_j - k_2 \right] x_i$$

and finally:

$$\dot{x}_i = \left(\sum_{j=0}^{N} n_{ij} x_j \right) x_i$$

where $n_{ij} = c\,(m_{ji} - k_1 m_{ij})$, for $j \neq 0$, $n_{0j} = 0$, for all j, $n_{i0} = -k_2$, for $i \neq 0$, and $x_0 = 1$ by convention.

Since this system of differential equations is not integrable using analytic methods, the algorithm uses the finite difference method to update the concentrations of the various antibody types. Therefore, the concentration of each antibody is approximated, on each step of the simulation, using the update rule:

$$x_i(t + h) = x_i(t) + h\dot{x}_i(t)$$

E.g.,

$$x_i(t + h) = x_i(t) + h\, x_i(t) \left(\sum_{j=0}^{N} n_{ij} x_j(t) \right)$$

If $k_1 = 1$ (equal stimulation and suppression) and $k_2 > 0$, then every antibody type eventually dies due to the damping term. Taking $k_1 < 1$ favors the formation of reaction loops, since all the numbers of a loop can gain concentration and thereby fight the damping term. As N increases, so do the length and number of loops. The robust properties of the loops allow the system to remember certain states even when the system is disturbed by the introduction of new types.

Along with the three parameters $(c, k_1$ and $k_2)$, which allow the tuning of the algorithm, two others are necessary for the specification of the algorithm. We need to set a *death threshold* and a *recruitment threshold*. The first establish the minimum concentration under what a given antibody type is eliminated from the population. The second indicates the minimum concentration above what a given antibody is sufficiently stimulated to initiate the recruitment process. Other parameters might be necessary depending on the recruitment process used on a given implementation.

We can now formulate the *Abstract Immune System Algorithm* as the following steps:

```
Randomly initialize initial population
Until termination condition is met do
    Update concentrations using equation formula
    Eliminate antibodies under death threshold
    Recruitment of new antibodies
```

As usual with this kind of algorithms, several types of termination conditions can be considered (time, number of generations, stability, maximum concentration, ...). As for the recruitment process, there are also many approaches that can be followed. The simplest one would be just to clone and mutate the antibodies with a probability proportional to their concentrations, but we can use a scheme where crossover and other evolutionary processes are involved.

4   Problems and Results

In order to use the algorithm to solve a given problem we need to know how to encode it and how to extract the result once the algorithm halts.

Each antibody represents a possible solution to the problem. Although the usual binary representation is always possible, sometimes it is preferable to use a different representation in order to avoid invalid solutions. These alternative representations are largely dependent on the problem at hand. Examples of these representations include: symbolic and tree-based representations.

Given an encoding, the second and probably the most important thing to do is the function that computes the affinity between one paratope and one epitope. For this function to be effective it must take into account not only the affinity between the two antibodies but also enhance this value according to the relative quality of the idiotop as a possible solution to the problem. This function corresponds to the m matrix.

A third thing that needs to be established is the method of recruitment used by the algorithm. The method can be as simple as clone and mutate antibodies with probability proportional to their concentrations, but it can also include other methods such as different types of crossover or any other evolutionary method. If the newly generated antibody doesn't exist in the population, then it is introduced with a fixed concentration. This is the way we express diversity.

As we might expect several possible encodings exist for a given problem and the choice will greatly influence the results obtained. Besides the different parameters, other factors that can influence the result are, for example, the initial population (although the algorithm states that it is randomly generated we can consider the situation where the initial population is carefully chosen) and its dimension (N).

The outcome of the algorithm will be the individual with the highest concentration of the population. Since the concentration is closely related with the probability of using a given antibody, it seems reasonable to take as answer to the problem the most probable potential answer found. Several other schemes could be considered depending on the problem itself.

5   The Probabilistic Computational Model

We consider probabilistic algorithms as in [2], chapter 6. Intuitively it is easy to recognize our Algorithm as probabilistic. The factors of uncertainty are the randomly generated initial population and the selection of antibodies for recruitment. Each of these factors can be described as probabilistic guesses. In this section we will address the problem of describing the Algorithm in a probabilistic format. First we will present its formal description and then we will analyze the probability related to each step of the process.

Consider a population of N antibody types. Each antibody is represented as a sequence of size n. In generation k, the bit j of the antibody i, is a random bit variable y_{ij}^k. The concentration of antibody i in generation k is the variable x_i^k, with $0 \leq x_i^k \leq 1$. We assume normalization: $\sum_{i=1}^N x_i^k = 1$, for all k.

In a rough analysis of our Algorithm, we introduce a few simplifying assumptions.

Assumption 1. *We assume a unique optimal solution so that each bit of the antibody type – solution has an unique correct value.*

Assumption 2. *We assume a surrogate affinity measure that is proportional to the antibody quality. The quality is measured by the number of correct bits in the antibody sequence.*

Although these assumptions are not always true, since many problems have multiple optima and the affinity measure can take any form, they are reasonable in many cases and hopefully will lead to new insights. By Assumption 1, there is a single correct value for each bit in the antibody type – solution. Let b_{ij}^k be the bit that tells us if y_{ij}^k is the correct value. Let $S_i^k = \sum_{j=1}^n b_{ij}^k$ be the number of correct values in antibody i in the generation k. By Assumption 2, the affinity function is proportional to S_i^k. The best element of the population is A^k that corresponds to $\max_i S_i^k$. Given a generation, with affinities $\{S_i^{k-1}\}_{1 \le i \le N}$, with $k > 0$, we can construct the next generation by successively updating and normalizing antibody concentrations, removing the antibody types with concentrations under the *death threshold* and incorporating newly generated antibodies. The *Abstract Immune System Algorithm* can be formally stated as follows:

(Initialize) Generate uniformly y_{ij}^0, for every $1 \le i \le N$, $1 \le j \le n$. Assign x_i^0 to its initial value. Calculate all S_i^0 and A^0. $k \to 1$.

(Iterate) While $A^{k-1} < n$ do

(Update) Use the rule

$$x_i^k = x_i^{k-1} + c \int_{k-1}^k x_i(t) \sum_{j=1}^N n_{ij} x_j(t) dt$$

to update the concentrations of the antibody types. Normalize the concentrations.

(Eliminate) Remove all antibody types with concentrations x_i^k less than the *death threshold*. Update the value of N.

(Recruitment) Randomly select for recruitment, with probabilities x_i^k, antibody types with concentrations greater than the *recruitment threshold*. Mutate the selected antibody types (modifying one bit of each) introducing the newly created antibodies into the population. Update the value of N.

(Evaluate) Evaluate each S_i^k. Determine A^k.

(Increment) $k \to k + 1$.

(end iteration)

Some steps of the Algorithm display a completely deterministic behaviour. *Update* will only affect the distribution of concentrations within the antibody types of the population. This will only influence the choice of an antibody for deletion and for recruitment. *Eliminate* will affect the population, but since an antibody eligible for recruitment could never be eligible for deletion it will not affect the *Recruitment* step. We will then concentrate in the two steps that display probabilistic behaviour.

For the initial population, y_{ij}^0 is generated uniformly. The probability of choosing a correct bit is always $1/2$ for each bit. Also b_{ij}^0 is like tossing n coins. By construction, y_{ij}^0 and b_{ij}^0 are independent variables. Hence S_i^0 has binomial distribution $B(n, 1/2)$. Further, all S_i^0 are independent from each other, though clearly not independent of the corresponding y_{ij}^0 and b_{ij}^0. The average affinity of the initial population is proportional to $E[S_i^0] = n/2$. With N independent members of the population, we have $P(A^0 \leq J) = P(\max_i S^0_i \leq J) = P(S^0_i \leq J, \forall i) = (F(J))^N$ where $F(J)$ is the cumulative distribution function of $B(m, 1/2)$, from 0 to J. The expectation value of A^0 is given by the expression

$$E[A^0] = \sum_{J=0}^{n-1} [1 - (F(J))^N].$$

In generation 0 the members of the population are independent. Subsequent generations depend on the previous ones. However, their correlation is greatly reduced by virtue of the distribution of affinities and for sufficiently large N.

The first part of the recruitment process is the selection of antibody types determined by the concentrations. Only antibody types with concentrations above the *recruitment threshold* are eligible for reproduction. The process of *Recruitment* will increase the size of the population. Given an antibody type p selected at random with affinity S^{k-1}_p, we derive the probability distribution of the affinity S^k_i of the offsprings. In this rough analysis we make one more assumption.

Assumption 3. *Given that the parent have J correct bits, the location of the correct bits is distributed uniformly and the correct bits are independent of each other.*

Assumption 3 will be true in early generations, but becomes less accurate as the population converges. The impact of this assumption will be judged empirically. Given the selected antibody type, the mutated bit will either change from an incorrect value to a correct one or change from a correct to an incorrect value. Then for the offspring we have

$$S^k_i = \sum_{j \neq l}^{n} b^k_{ij} + w_l$$

where w_l is a Bernoulli variable corresponding to probability $1/2$. The maximum value of J can take is given when all the correct bits remain unchanged and the bit selected mutates to a correct value. In this situation we have $S^k_i = S^{k-1}_i + 1$. Under the Assumption 3 we have

$$P(S^k_i = J | S^{k-1}_p = J') = \binom{n}{J}(\frac{J'}{n})^J (1 - \frac{J'}{n})^{(n-J)}$$

For a sufficiently large N, we may approximate $P(S^k_i \leq J_i, 1 \leq i \leq N) \approx \prod_{i=1}^{N} P(S^k_i \leq J_i)$. According to our assumptions, we have

$$P(S^k_i \leq J) \approx \sum_{I=0}^{n} P(S^k_i \leq J | S^{k-1}_p = I) P(S^{k-1}_p = I)$$

The first factor of the sum can be computed from the equation above for $P(S^k_i = J | S^{k-1}_p = J')$. The second factor is given as the inductive step. Then

$$E[S^k_i] = \sum_{J=0}^{n-1} [1 - P(S^k_i \leq J)^N].$$

Since $P(A^k \leq J) \approx [P(S_p^k \leq J)]^N$, we get $E[A^k] \approx \sum_{J=0}^{n-1}[1 - P(A^k \leq J)]$. To compute $E[A^k]$, we compute the anchor distribution of S_i^0 as given before, and then, inductively, we compute the distributions of S_i^k.

6 Examples

In this section we present two of the examples used to test the Algorithm. A workbench of classes/interfaces in Java was programmed. The source code, its documentation and the examples (to be tested by the reader) can be found at the address http://ctp.di.fct.unl.pt/~jddp/immune/. The default settings of the parameters of the workbench correspond the values that provided best experimental results.

The first example selected is a classical maximization problem. The aim of this example is to show how to use the Algorithm in its simplest form to solve a problem that is very well studied for other biologically inspired algorithms. We considered the power function $f(x) = (x/c)^n$, where c is a normalizing constant (with $n = 10$).

We considered a binary representation with 30 bits per antibody corresponding to a large search space. Then we had to define an affinity function. As suggested in Section 4, we considered two factors to compute this function. The first factor represents the quality of the antibody, given by the function f we aim to optimize. The second factor represents the correlation between an antibody and any other antibody of the population, for this we used the Hamming distance between the two antibody strings u and v of size m

$$\mathcal{H}(u,v) = \sum_{i=1}^{m} u_i\,(1 - v_i) + \sum_{i=1}^{m} v_i\,(1 - u_i).$$

The final affinity function is

$$\textit{Affinity}(u,v) = f(u) \times \mathcal{H}(u,v)$$

The third aspect is the recruitment method. We add to the population mutated clones of the antibodies with concentration above the *recruitment threshold*.

The Algorithm was used with different initial population sizes and various parameters and, nine out of ten times, it was capable of finding a near optimal solution in less than a hundred generations.

As second example we considered the *Iterated Prisoner's Dilemma*. The aim of this example is to show the use of the Algorithm with a less trivial problem, and the use of more sophisticated antibody representation and affinity function.

The prisoner's dilemma is a classic problem of conflict and cooperation [20,1]. In its simplest form each of two players has a choice of cooperating with the other or defecting. Depending on the two players' decisions, each receives payoff according to a payoff matrix. When both players cooperate they are both rewarded at an equal, intermediate level (reward, R). When only one player defects, he receives the highest level of payoff (temptation, T), while the other player gets the sucker's just deserts (sucker, S). When both players defect they each receive an intermediate penalty (P). The prisoner's dilemma has often been cited as a simple yet realistic model of the inherent difficulty of achieving cooperative behaviour when rewards are available for the successful miscreant. The problem is called the prisoner's dilemma because it is an abstraction of the situation felt by a prisoner who can either cut a deal with the prosecutor and thereby rat on his partner in crime (defect) or keep silent and therefore tell nothing of the misdeed (cooperate).

Table 1. Prisoner's Dilemma Payoff Matrix

	Player 2	
	Cooperate	Defect
Player 1 Cooperate	R=6, R=6	S=0, T=10
Defect	T=10, S=0	P=2, P=2

The problem is made more interesting by playing it repeatedly with the same group of players, thereby permitting partial time histories of behaviour to guide future decisions. This so-called iterated prisoner's dilemma has drawn interest from game theorists for a while.

A strategy for the Iterated Prisoner's Dilemma is not obvious. The goal to the algorithm was to come up with the best possible strategy.

As ever, our first concern is to provide a representation for the possible solutions of the problem. To represent a strategy we will need to decide what to do next based on the history of past plays.

In our representation, to decide on a current move, a current player will look at the past three rounds before making a decision. If a single round can be represented by 2 bits: CC, CD, DC, DD, a three round memory requires 6 bits. By choosing to represent a cooperation (C) as a 0 and a defect (D) as a 1, a simple history pattern can be recorded. For instance, 10 00 01 will represent that Player 2 defected three rounds ago, they both cooperated two rounds ago, and Player 1 defected last round. In order to choose a current move, a Player must have a strategy for all possible previous three rounds. In other words, a Player must have 64 (2^6) different strategies on how to play. Also, a strategy consists of defecting or cooperating, which can be represented by a single bit (0 for cooperate and 1 for defect). These facts allow the creation of our bit string for the algorithm. Each bit represents a strategy for a given past three rounds. The index value for the bit string is directly computable from the 6-bit history, it is merely the conversion of the 6-bit history into the corresponding number. So, the representation of a strategy for the prisoner's dilemma is a 70-bit string. The first 64-bits are to be used to determine the current move based on past moves. The last 6 bits are to be used to determine the first move. Of course this representation can be generalized to consider the n previous rounds to make the decision, instead of just three. Note however that the size of the antibodies will increase exponentially.

Our second concern is to choose an affinity function. The obvious choice is to make different strategies play against each other and use the difference of the scores as the affinity. Obviously, the number of rounds we choose to play will influence the accuracy of the affinity function. On the other hand, since the algorithm will test the affinity of each antibody against all other, choosing a large number of rounds will slow down the algorithm greatly. For this example we used an affinity function that will play 20 rounds.

Our last concern is to provide a recruitment method for the algorithm. As we did on the first example, as recruitment method we use cloning and mutation. A mutation function can easily be implemented by randomly changing the value of a small number of bits.

Finally, as a way to improve the performance of the algorithm, we introduced in the initial population several antibodies representing well known proposed strategies, such as:

- **nice** : player cooperates all the time;
- **mean** : player defects all the time;
- **tit-for-tat** : player cooperates for the first move, and then copies the other player's last move;
- **opp-tit-for-tat** : player defects for the first move, and then copies the other player's last move;
- **cd** : periodic behavior, plays $CDCDCD$
- **ccd** : periodic behavior, plays $CCDCCD$
- **dcc** : periodic behavior, plays $DDCDDC$
- **pavlov** : player cooperates if both players played the same action in the last round.

The introduction of several well known strategies in the initial population had the objective of providing the algorithm with some valid and efficient competition. Experimental results show that the algorithm provided best, or at least fastest results, when this was done.

Nevertheless, even with a random initial population the algorithm was able to discover strategies that beat the overall performance of the **tit-for-tat** strategy.

7 Conclusions and Future Work

We proposed a problem solving algorithm based on the model of the immune system of Farmer *et al*, referred to as *Abstract Immune System Algorithm*. The Algorithm proved to be capable to solve small to moderate size problems in an efficient and reliable way.

The proposed algorithm falls under the category of biologically inspired soft-computing systems. The following table (taken from [17]) summarizes the relation between the three selectionist theories, mentioned in our work, namely (1) the theory of (natural) evolution of species (New-Darwinism) [left]; (2) the clonal selection theory (Jerne's Clonal Selection theory) [middle]; (3) the theory of neuronal group selection (Edelman's Neural Darwinism) [right]:

	Natural evolution	Immune system	Brain
Selectionist theory	New-Darwinism	Clonal selection theory	Neuronal group
Generator of diversity	Mutation and crossover	Recruitment strategy	Cell movement, differentiation and death
Unity of selection	Genotype / phenotype	Antibody type	Neuronal group
Amplification process	Reproduction of the fitter individuals	Secretion of free antibodies and reproduction of the corresponding B-lymphocytes	Modification of the synaptic weights of the connections
Explains	Evolution	Functioning	Functioning
Adaptation	Innovation	Innovation	Adjustment

Future work might concentrate on the study of variants of the Algorithm and the impact of its parameters. As with evolutionary algorithms, the *Abstract Immune System Algorithm* can be subject to variations, concerning either the order of operations or different recruitment strategies. Another area for further exploration is the experimentation of the algorithm in a larger set of problems and the study of its adequacy. We might be concerned both with the quality of the results and the characterization of the class of problems best suited for the algorithm. The consideration of different representations, affinity functions and recruitment methods dealing with well known problems can prove of particular interest.

Acknowledgements. The authors are indebted to Jorge Orestes Cerdeira, Carlos Lourenço and Helder Coelho for detailed criticism about this work. José Félix Costa also thanks to his previous students João Maças and Francisco Martins who started in 1995 this project with him.

References

1. Axelrod, R.: The Complexity of Cooperation: Agent-Based Models of Competition and Collaboration. Princeton Press, Princeton, NJ (1997)
2. Balcázar, J., Díaz, J., Gabarró, J.: Structural Complexity I, 2nd edn. Springer, Heidelberg (1995)
3. Bersini, H.: Reinforcement learning and recruitment mechanism for adaptive distributed control. Technical Report IR/IRIDIA/92-4, Institut de Recherches Interdisciplinaires et de Développements en Intelligence Artificielle (1992)
4. Bersini, H., Seront, G.: Optimizing with the immune recruitment mechanism. Technical Report TR/IRIDIA/93-6, Institut de Recherches Interdisciplinaires et de Développements en Intelligence Artificielle (1993)

5. Bersini, H., Varela, F.J.: The immune recruitment mechanism: A selective evolutionary strategy. Technical Report IR/IRIDIA/91-5, Institut de Recherches Interdisciplinaires et de Développements en Intelligence Artificielle (1991)
6. Burnet, F.M.: The Clonal Selection Theory of Acquired Immunity. Cambridge University Press, Cambridge (1959)
7. Celada, F., Seiden, P.E.: A computer model of cellular interactions in the immune system. Immunology Today 13(2), 56–62 (1992)
8. Coutinho, A.: The network theory: 21 years later. Scandinavian Journal of Immunology 42, 3–8 (1995)
9. DasGupta, D. (ed.): Artficial Immune Systems and Their Applications. Springer, Heidelberg (1998)
10. de Castro, L.N, Timmis, J.: Artificial Immune Systems: A New Computational Approach. Springer, Heidelberg (2002)
11. Farmer, J.D., Packard, N.H., Perelson, A.S.: The immune system, adaptation, and machine learning. Physica D 22, 187–204 (1986)
12. Forsdyke, D.R.: The origins of the clonal selection theory of immunity as a case study for evaluation in science. The FASEB Journal 9, 164–166 (1995)
13. Herbert, W.J., Wilkinson, P.C., Stott, D.I: The Dictionary of Immunology, 4th edn. Academic Press, San Diego (1995)
14. Jerne, N.K.: The natural-selection theory of antibody formation. Proceedings of The National Academy of Sciences 41, 849–856 (1955)
15. Jerne, N.K.: Towards a network theory of the immune system. Ann. Immunol. (Inst. Pasteur) 125(C), 373–389 (1974)
16. Lord, C.: An emergent homeostatic model of immunological memory. Technical report, Carnegie Mellon Information Networking Institute (February 2002)
17. Manderick, B.: The importance of selectionist systems for cognition. In: Patton, R. (ed.) Computing with Biological Metaphors, pp. 373–392. Chapman and Hall, Sydney, Australia (1994)
18. Paul, W.E. (ed.): Fundamental Immunology, 3rd edn. Raven Press, New York (1993)
19. Perelson, A.S., Weisbuch, G.: Immunology for physicists. Reviews of Modern Physics 69(4), 1219–1263 (1997)
20. Poundstone, W.: Prisoner's Dilemma: John von Neumann, Game Theory and the Puzzle of the Bomb. Doubleday Publishing (1993)
21. Tarakanov, A.O., Skormin, V.A., Sokolova, S.P.: Immunocomputing: Principles and Applications. Springer, Heidelberg (2003)

Taming Non-compositionality Using New Binders

Frédéric Prost

LIG

46, av Félix Viallet, F-38031 Grenoble, France

`Frederic.Prost@imag.fr`

Abstract. We propose an extension of the traditional λ-calculus in which terms are used to control an outside computing device (quantum computer, DNA computer...). We introduce two new binders: ν and ρ. In $\nu x.M$, x denotes an abstract resource of the outside computing device, whereas in $\rho x.M$, x denotes a concrete resource. These two binders have different properties (in terms of α-conversion, scope extrusion, convertibility) than the ones of standard λ-binder. We illustrate the potential benefits of our approach with a study of a quantum computing language in which these new binders prove meaningful. We introduce a typing system for this quantum computing framework in which linearity is only required for concrete quantum bits offering a greater expressiveness than previous propositions.

1 Introduction

λ-calculus [8] is a particularly well suited formalism to study functional programs. Any λ-variable can be replaced by any λ-term in the pure λ-calculus. λ-abstraction can be seen in the following way: the term $\lambda x.M$ represents an algorithm M in which x denotes another completely abstracted algorithm, thus x may be seen as a black box. This approach is deeply compositional. The "meaning" of M can be precisely defined with relation to the "meaning" of x. This inner compositionality of λ-calculus is reflected by the success of type systems for λ-caculi [9]. Those type systems can be used with good effect to statically analyze programs (e.g. [2,20,19]).

Some λ-calculus variants have been developed to study functional programing languages including imperative features like ML (e.g. [3,14,18]). The idea is to introduce a global state and to refer to it via variable names. Name management is handled via standard λ-abstraction mechanism. Typically in ML-like languages [16] one can define a variable of type int by P $\overset{def}{=}$ let x=ref 0 in M which is syntactic sugar for $(\lambda x.M \ (\texttt{new } 0))$, where new is a side-effect function that adds a new cell referred by x and whose value is 0 in the global state. α-conversion of the λ-abstraction insures hygienic properties of such computing models, it handles the fresh-name generation problem. For instance consider:

$$(\lambda y.\langle y, y \rangle \ P)$$

S.G. Akl et al.(Eds.): UC 2007, LNCS 4618, pp. 150–162, 2007.

where P is duplicated. The actual evaluation of this term will have to handle the fact that x cannot refer to two different cells. From a a theoretical point of view this problem is handled by α-conversion.

Introducing a global state enhances the expressiveness of the language but it has a cost: compositionality is lost. Indeed, side effects on the global state lead to non compositional behavior of programs. Consider the following program:

$$Q \stackrel{def}{=} \lambda y.(x := !x + 1; y := !y + 1)$$

where $!x$ stands for the content of an imperative cell x. It is clear that $(Q\ x)$ increments the content of x by 2 whereas $(Q\ z)$, provided that $x \neq z$, increments the content of x by 1. The problems get more complicated if you consider a language in which variables can be references, and thus with potential aliasing of variables: if z is aliased to x (but is syntacticly different from x) then $(Q\ z)$ increments x by 2. Static analysis of such programs is not an easy task, it requires sophisticated approach very different from usual typing systems (see [11]).

Things can get even more complicated when the global state is not classical. If you consider quantum computing [17], the global state is an array of quantum bits. Several functional programming language have been proposed for quantum computation (e.g. [22,21,6]). In a quantum functional programing language, quantum bits can be entangled by side effects that go beyond the lexical scope of their definitions. and thus such programing languages exhibit a highly-non compositional behavior. Consider the program

$$Q \stackrel{def}{=} \lambda x, y.(\mathfrak{C}\mathbf{not}\ \ < x, y >)$$

Q executes the conditional not on two quantum bits x, y. Therefore, $(P\ q_1\ q_2)$ may entangle qbits q_1 and q_2. But if q_1 is entangled to q_0, and q_2 is entangled to q_3, then $(P\ q_1\ q_2)$ also entangles q_0 to q_3 ! This simplistic example shows that the classical binder, λ, is not well suited to deal with quantum bits and more generally with global references.

In this paper we propose to extend the λ-calculus by adding two new binders: ν and ρ. These binders exhibit different fundamental properties than the ones of the λ binder. We show that these binders are more adapted for the design of a functional programming language including a global state. νx declares a new resource that can be used. It is very close to the ν of the π-calculus [15] (hence the notation), but has not the scope extrusion property. In $\nu x.M$, x is an abstract reference: x refers actually to nothing in the global state. On the other hand ρ binder defines a concrete reference to something actual in the global state. Because of possible side effects, ρ has the full scope extrusion property, indeed operations on a local variable may have implications outside the scope of its definition as explained earlier.

In section 2 we introduce λ_{GS}, a λ-calculus based framework to deal with global states. We define a specific λ_{GS} calculus for quantum computation: λ_{GS}^Q in section 3. We give a type system for λ_{GS}^Q allowing greater programming flexibility than previous propositions of the literature. Finally, we conclude and discuss future works in section 4.

Knowledge of basics of quantum computation is supposed. We refer to [7,1] for quantum computing tutorials.

2   Global State Calculus

2.1 λ_{GS} Ideas

A λ_{GS} calculus is defined relatively to an outside computing device on which computations/actions can be performed (such as a quantum state, DNA computer etc.). Those actions controlled via λ_{GS} terms. λ_{GS} calculus may be viewed as a system providing a functional control of a computing device. Thus a program in λ_{GS} calculus has two parts: a hard one (the computing device viewed as a global state which is a physical artifact used to compute) and a soft one (the functional term that controls the computing process).

The functional part of λ_{GS} is implemented by standard λ-calculus terms with extra features to deal with the global state. Interactions with the global state are handled by two mechanisms:

1. Locations: from the term point of view, locations are just names. From the global state point of view, locations are the addresses of physical entities (the actual address of information stored in RAM, a specific quantum bit etc.).
2. Actions: from the term point of view, actions are terms that can be evaluated. The evaluation depend on the global state as well as on the parameters of the action (for instance the result of the measurement of a quantum bit). From the global state point of view, actions perform actual modifications of it (update of a RAM cell, application of a unary gate to quantum bits).

2.2 λ_{GS} Definition

A λ_{GS} framework is defined up to the choice of a global state and the according actions.

Terms

Definition 1 (Terms). *The set of terms $\mathcal{T}$ is inductively defined as follows:*

$$
\begin{aligned}
M, N, P ::= {}& x \mid \ell \mid \lambda x.M \mid (M \ \ N) \\
& \mid \langle M, N \rangle \mid \text{let } \langle x, y \rangle = M \text{ in } N \\
& \mid \nu x.M \mid \rho\ell.M \mid !M \\
& \mid \mathsf{a}_i(M)
\end{aligned}
$$

where x, ℓ are respectively variable and location names defined over a countable set $\mathcal{V}$. $\mathcal{L}$ is the subset of $\mathcal{V}$ denoting locations.

There is no formal distinction between x and ℓ. However we use the following convention: ℓ is used to denote a location (that is a reference to the outside global state) whereas x denotes a standard λ-variable.

Contexts are defined as usual : a context $C[\cdot]$ is a term where $\cdot \in \mathcal{V}$ occurs exactly once. By $C[M]$ we denote the syntactical replacement of $\cdot$ with M (with name capture).

The first two lines of definition 1 define standard pure λ-terms with tuples ($\overrightarrow{M}$ denotes tuples of terms : $\overrightarrow{M} \overset{def}{=} \langle M_1, \langle M_2, \langle \ldots, M_n \rangle \ldots \rangle\rangle$). Tuples are native in λ_{GS} because of actions, indeed actions cannot be curryfied: they are not functions in a λ-calculus sense.

In λ_{GS}, there are three binders λ, ν and ρ. Thus, we have the following definition of free variables.

Definition 2 (Free variables). *Free variables are defined by:*

$$FV(x) = \{x\} \quad FV(!M) = FV(\mathsf{a}_i(M)) = FV(M)$$

$$FV(\lambda x.M) = FV(\nu x.M) = FV(\rho x.M) = FV(M) \setminus \{x\}$$

$$FV((M \ N)) = FV(\langle M, N \rangle) = FV(M) \cup FV(N)$$

$$FV(\mathsf{let} \ \langle x, y \rangle = M \ \mathsf{in} \ N) = (FV(M) \cup FV(N)) \setminus \{x, y\}$$

In $\nu x.M$, x is an abstract fresh location and is bound in M. The idea, is that $\nu x.M$ is a piece of code that will use a global state information referred by x. On the other hand $\rho \ell.M$ denotes a piece of code that *actually* uses a concrete global state information. ℓ is bound in M in $\rho \ell.M$. As usual we write $\rho \ell_1, \ldots, \ell_n.M$ as shorthand for $\rho \ell_1.\rho \ell_2.\ldots.\rho \ell_n.M$.

$!M$ is the *realization* action. It transforms an abstract location into a concrete one. Intuitively (formal definition follows) we have the reduction:

$$C[!\nu x.M] \rightarrow C[\rho \ell.M[x := \ell]]$$

where ℓ is a fresh location relatively to $FV(C[!\nu x.M])$. The idea is that realization has a side effect on the global state: it actually creates a new resource.

a_i are actions and are relative to the global state. For example, if the global state stores terms actions are affectation and pointer dereferencing. If the state is made of quantum bits, then typical actions include unary gates (e.g. phase, Conditional Not, Hadamard) and measurement.

Equivalences. We now define several equivalence relations over terms. Those relations exhibit differences between the three λ_{GS} binders.

Definition 3 (Term equivalence). *We have the following equivalences between terms:*

− λ *and* ν *binders generate* α-*equivalence:*

$$\lambda x.M =_\alpha \lambda z.M[x := z]$$
$$\nu x.M =_\alpha \nu z.M[x := z]$$

with $z \notin FV(M)$

— ν and ρ binders generate commutation equivalence:

$$\rho\ell.\rho\ell'.M =_\xi \rho\ell'.\rho\ell.M$$
$$\nu x.\nu y.M =_\xi \nu y.\nu x.M$$

— ρ binder generates scope extrusion equivalence:

$$C[\rho\ell.M] =_\sigma \rho\ell.C[M]$$

for any context $C[\cdot]$.

— ρ binder generates garbage collection equivalence:

$$\rho o.M =_\gamma M \quad o \notin M$$

Let $\doteq$ be the union of $=_\alpha$, $=_\sigma$, $=_\xi$ and $=_\gamma$.

Scope extrusion is the main λ_{GS} mechanism to handle non-compositionality. Indeed, ℓ in $\rho\ell.M$ refers to some outside computing device, thus any action on ℓ can have side effects that go beyond the lexical scope of $\rho\ell.M$. Consider for instance a quantum computation model where locations refers to quantum bits. In the following term:

$$N \equiv \rho\ell.\langle\rho\ell_1.\rho\ell_2.M, \rho\ell_3.\mathfrak{Cnot}(\langle\ell_1, \ell_3\rangle)\rangle$$

where $\mathfrak{Cnot}$ denotes the conditional not, the problem is the following : if ℓ, ℓ_1 and ℓ_2 are entangled in M, then a conditional not on ℓ, ℓ_3 can entangle ℓ_3 with ℓ_1, ℓ_2. The idea is that once a concrete physical object is defined in a term then this object may influence, or it may be influenced, by any other part of the surrounding context. The standard form of N (see def. 1 below) is:

$$\rho\ell, \ell_1, \ell_2, \ell_3.\langle M, \mathfrak{Cnot}(\langle\ell_1, \ell_3\rangle)\rangle$$

The idea is that scope extrusion is a syntactic mean to take into account non compositionality. Indeed, in the standard form of N, the range of ℓ_3 is extended in such a way that it can interact, or more precisely that it is possible to describe within the term the interactions, with ℓ_2 for instance.

On the other hand ν binder only refer to an abstract pointer. It is an abstraction over outside computing devices that are going to be used to perform some computation. Consider the following program:

$$M1 = (\lambda x.\langle x, x\rangle \quad \nu y.y)$$

we don't want this to be equivalent to

$$M2 = \nu y.(\lambda x.\langle x, x\rangle \quad y)$$

since the evaluation of $!M1$ creates two different resources, whereas $M2$ only creates one resource.

Following these equivalences, a notion of standard form for terms naturally emerges (it corresponds to a distinguished term for each $\doteq$ equivalence class).

Fact 1 (Standard form). *For any term M, there exists $N \doteq M$ such that $N \equiv \rho\ell_1.\rho\ell_2.\dots.\rho\ell_n.N_\rho$ with $\{\ell_1, \dots, \ell_n\} \subseteq FV(N_\rho)$ and N_ρ has no subterm of the form $\rho\ell.N'$.*

In the following we work modulo $\doteq$. That is we suppose that terms are in standard form. Thus, when we consider a term t having a form different from $\rho\ell.M$, it implicitly means that t does not contain any ρ binder. Moreover $\rho\ell.M$ also implicitly means that ℓ occurs in M.

Operational semantics. The global state is a mathematical view of a computing device. Let $\mathcal{S}$ be a function from $I\!N$ to some domain $\mathcal{D}$ which denotes the actual physical objects concerned (DNA, quantum bits etc). We write $\emptyset$ the state nowhere defined. The bridge between λ_{GS} terms and global state is done using a Linking function that maps locations (λ_{GS} terms) to naturals (the address of the physical entity). For instance in a quantum λ_{GS} system, the computing device can be an array of quantum bits. A location ℓ is associated to some qbits ($\mathcal{K}(\ell) = 12$ indicates that ℓ is the 12th qbit of the array of quantum bits).

A λ_{GS} system is defined up to a family of *actions* that can be applied to the global state. Take for instance a usual computer memory, classically you have three actions to manipulate it: you can allocate a new memory cell to store an information, read the content of a memory cell and update the content of an existing memory cell.

Each action $\mathbf{a}_i$, has a side effect on the computing device and yields a value that can be used for further computations. Thus, if the arity of $\mathbf{a}_i$ is n_i we must provide two functions :

$$\mathcal{F}_i^{\mathcal{D},\mathcal{K}} : (\mathcal{T}^{n_i} \times (I\!N \to \mathcal{D})) \to (I\!N \to \mathcal{D})$$
$$\mathcal{F}_i^{\mathcal{T},\mathcal{K}} : \mathcal{T}^{n_i} \to \mathcal{T}$$

$\mathcal{F}_i^{\mathcal{D},\mathcal{K}}$ models the side effect on the computing device by modifying the global state (modification of $\mathcal{S}$) and $\mathcal{F}_i^{\mathcal{T},\mathcal{K}}$ denotes the value returned to the λ_{GS} term.

As usual if f is a function we denote by $f \uplus \{x \mapsto v\}$ the function g such that $g(y) = f(y)$ for all $y \neq x$ and $g(x) = v$.

The evaluation of pure terms is done via β-reduction as in the standard λ-calculus. We write $M[x := N]$ the term M where all free occurrences of x have been replaced with N.

Definition 4 (Functional evaluation)

$$(\lambda x.M \ \ N) \to_\beta M[x := N]$$

λ_{GS} evaluation process may alternate evaluation on pure terms as well as actions on the global state. We have the following definition of the computation:

Definition 5 (λ_{GS} computation). *We define $\leadsto$, the computation relation, between tuples $[\mathcal{S}, \mathcal{K}, M]$*

– *Function: if $M \to_\beta M'$:*

$$[\mathcal{S}, \mathcal{K}, M] \rightsquigarrow [\mathcal{S}, \mathcal{K}, M']$$

– *Action:*

$$[\mathcal{S}, \mathcal{K}, (a_i \ \overrightarrow{M})] \rightsquigarrow [\mathcal{F}_i^{\mathcal{D},\mathcal{K}}(\overrightarrow{M}, \mathcal{S}), \mathcal{K}, \mathcal{F}_i^{\mathcal{T},\mathcal{K}}(\overrightarrow{M})]$$

– *Realization:*

$$[\mathcal{S}, \mathcal{K}, !\nu x.M] \rightsquigarrow [\mathcal{S} \uplus \{n+1 \mapsto d\}, \mathcal{K} \uplus \{o \mapsto n+1\}, \rho o.(M[x := o])]$$

with o a fresh name, d a distinguished value of $\mathcal{D}$ and $\mathcal{S}$ defined on $\{1, \ldots, n\}$.

2.3 λ_{GS} Properties

Due to imperative nature of global state, it is clear that λ_{GS} might not be a confluent calculus. Actually λ_{GS} is intrinsically non confluent even with the simplest domain and actions. Consider the unit $\mathcal{D} = \{\mathfrak{d}\}$, and λ_{GS} with no actions. Now consider the following computation where β-reduction and realizations are alternatively performed. First, lets reduce the β-redex:

$$
\begin{aligned}
[\emptyset, \emptyset, (\lambda x.(x \ \ x) \ \ !\nu y.y)] &\rightsquigarrow [\emptyset, \emptyset, (!\nu y.y \ \ !\nu y.y)] \\
&\rightsquigarrow [\{1 \mapsto \mathfrak{d}\}, \{o \mapsto 1\}, \rho o.(!\nu y.y \ \ o)] \\
&\rightsquigarrow [\{1 \mapsto \mathfrak{d}, 2 \mapsto \mathfrak{d}\}, \{o \mapsto 1, o' \mapsto 2\}, \rho o'.\rho o.(o' \ \ o)]
\end{aligned}
$$

Second, lets reduce the κ-redex:

$$
\begin{aligned}
[\emptyset, \emptyset, (\lambda x.(x \ \ x) \ \ \nu y.y)] &\rightsquigarrow [\{1 \mapsto \mathfrak{d}\}, \{o \mapsto 1\}, \rho o.(\lambda x.(x \ \ x) \ \ o)] \\
&\rightsquigarrow [\{1 \mapsto \mathfrak{d}\}, \{o \mapsto 1\}, \rho o.(o \ \ o)]
\end{aligned}
$$

Thus, in order to have confluence, on has to fix a reduction strategy. We will not discuss this point further here, and let the study of reduction strategies for future work.

3 λ_{GS} for Quantum Computing

In this section we show how λ_{GS} can be used to define a functional quantum programming language. We develop a typing system which insures the "no-cloning" theorem. Our typing much more flexible than previous propositions based on a strict linear typing discipline for quantum bits (e.g. [21]). This typing system relies heavily on the new binders that make a syntactical distinction between abstract and concrete quantum bits.

3.1 λ_{GS}^Q Definition

λ_{GS}^Q is a quantum programing system based on λ_{GS}. We suppose that the reader has basic knowledges of quantum computing (see [17]). It is a functional programming language for quantum computers based on the QRAM model [13]. The physical device of λ_{GS}^Q is an array of qbits. n qbits are represented as normalized vector of the Hilbert space $\otimes_{i=1}^n C^2$ noted as a ket vector : $|\varphi\rangle$.

quantum actions are standard unitary quantum gates like $\mathfrak{Cnot}, \mathfrak{T}, \mathfrak{H}$ (respectively control not, phase and Hadamard gates see [17] for precise definition) and measure $\mathfrak{M}$. λ_{GS}^Q requires two boolean constants 0 and 1 corresponding to the result of measurement.

Quantum measurement is probabilistic, therefore the measure action $\mathfrak{M}$ is probabilistic. However we do not focus on this particular point in this paper: we are just interested in developing a typing system for λ_{GS}^Q. Functions defining unitary actions are as follow:

$$\mathcal{F}_{\mathfrak{T}}^{|\varphi\rangle,\mathcal{K}}(q) = \mathfrak{T}_{\mathcal{K}_q}(|\varphi\rangle)$$

$$\mathcal{F}_{\mathfrak{H}}^{|\varphi\rangle,\mathcal{K}}(q) = \mathfrak{Cnot}_{\mathcal{K}_q}(|\varphi\rangle) \qquad \mathcal{F}_{\mathfrak{Cnot}}^{|\varphi\rangle,\mathcal{K}}(\langle q,q'\rangle) = \mathfrak{H}_{\mathcal{K}_q,\mathcal{K}(q')}(|\varphi\rangle)$$

$$\mathcal{F}_{\mathfrak{T}}^{\mathcal{T},\mathcal{K}}(q) = \mathcal{F}_{\mathfrak{H}}^{\mathcal{T},\mathcal{K}}(q) = q \qquad \mathcal{F}_{\mathfrak{Cnot}}^{\mathcal{T},\mathcal{K}}(\langle q,q'\rangle) = \langle q,q'\rangle$$

where $\mathfrak{T}_i, \mathfrak{H}_i, \mathfrak{Cnot}_{i,j}$ are the respective phase, hadamard and conditional not gates on quantum bits i and j (see [17]).

Let $|\varphi\rangle = \alpha |\varphi_O\rangle + \beta |\varphi_1\rangle$ be normalized with $|0\rangle$ and $|1\rangle$ being the i-th quantum bit and

$$|\varphi_O\rangle = \Sigma_i \alpha_i |\phi_i^0\rangle \otimes |0\rangle \otimes |\psi_i^0\rangle \qquad |\varphi_1\rangle = \Sigma_i \beta_i |\phi_i^1\rangle \otimes |1\rangle \otimes |\psi_i^1\rangle$$

then we define $\mu_0 = |\alpha|^2$ and $\mu_1 = |\beta|^2$. We use $=_p$ to define probabilistic functions: $f(x) =_p y$ means that $f(x)$ yields y with probability p.

$$\mathcal{F}_{\mathfrak{T}}^{\alpha|\varphi_O\rangle+\beta|\varphi_1\rangle,\mathcal{K}}(q) =_{\mu_0} 0 \qquad \mathcal{F}_{\mathfrak{T}}^{\alpha|\varphi_O\rangle+\beta|\varphi_1\rangle,,\mathcal{K}}(q) =_{\mu_1} 1$$

$$\mathcal{F}_{\mathfrak{T}}^{\alpha\mathcal{D}_O+\beta|\varphi_1\rangle,\mathcal{K}}(q) = \alpha|\varphi_O\rangle \qquad \mathcal{F}_{\mathfrak{T}}^{\alpha|\varphi_O\rangle+\beta|\varphi_1\rangle,\mathcal{K}}(q) = \beta|\varphi_1\rangle$$

Realization is implemented by choosing $|0\rangle$ as default value, hence:

$$[|\varphi\rangle,\mathcal{K},!\nu x.M] \rightsquigarrow [\mathcal{S}\otimes|0\rangle,\mathcal{K} \uplus \{o \mapsto n+1\},\rho o.(M[x := o])]$$

3.2 λ_{GS}^Q Typing System

We now define a typing judgment for λ_{GS}^Q that ensures the no-cloning property of quantum states through a linear typing discipline.

Definition 6 (λ_{GS}^Q types). *Types are defined by:*

$$\tau ::= \mathbf{B} \mid \mathbf{Q} \mid \tau \to \tau \mid \tau \times \tau \mid \Diamond\tau$$

For the sake of simplicity we only consider two base types: $\mathbf{B}$ is the type of bits having $0, 1$ as constants, and $\mathbf{Q}$ is the type of quantum bits. Note that there are no constants of type $\mathbf{Q}$ in λ_{GS}^Q. Quantum bits are manipulated only through location names and linking function. It is because of the non-locality of quantum computations that quantum bits cannot be directly manipulated at the level of terms (see [21] for further explanations). Unlike [21] there is no primitive notion of linear types (there is no linear application function as well as exponential types): linearity is handled through ν and ρ binders only. Type $\Diamond\tau$ is used to identify ν redexes. Arrow and product types have their usual meanings.

In λ_{GS}^Q, ν and ρ only binds quantum bits locations. Thus x, ℓ respectively in $\nu x.N$ and $\rho\ell.M$ denote a location of a respectively abstract and concrete quantum bit and must be typed with $\mathbf{Q}$.

Typing contexts, written $\Gamma; \Delta$, are lists of couples made of variables and types in which variables are uniquely defined. In order to help the reading we cut typing contexts in two parts, the first part (Γ) is classical whereas the second one (Δ) is linear. For instance in $x_1 : \tau_1, \ldots, x_n : \tau_n; x_1 : \tau_1, \ldots, x_m : \tau_m$, each x_i is classical and can be reused (contraction rule $[CTN]$) and discarded as will (weakening rule $[\mathrm{WkgC}]$),and each y_i is linear and can be used exactly once but can be discarded (weakening rule $[\mathrm{WkgL}]$).

It appears that linking function and global state do not play any role in the typing process since locations can only contains quantum bits. Therefore typing judgment is simply defined on λ_{GS} terms (and not in tuples including a global state a linking function and a term).

Definition 7 (Typing judgement). λ_{GS}^Q *typing judgment is a tuple* $\Gamma; \Delta \vdash M : \tau$, *where* Γ, Δ *are typing contexts,* M *is a* λ_{GS}^Q *term and* τ *is a* λ_{GS}^Q *type. Typing rules are given in Fig. 1.*

Note that when $.;. \vdash M : \tau$ implicitly means (remember our convention to consider terms in standard form), that ρ does not occur in M. In facts typing rules are defined on standard forms. Rule $[\rho I]$ is only applicable once at the root of a typing tree and removes all ρ of the term.

λ_{GS}^Q typing systems is both classical (relatively to typing context Γ) and linear (relatively to typing context Δ). The idea being that quantum data have to be treated in a linear way (because of the no cloning property) whereas classical data can be arbitrarily copied. Rules $[AxC], [WkgC], [ExC], [\to IC], [\times EC]$ are the classical counterparts of linear rules $[AxL], [WkgL], [ExL], [\to IL], [\times EL]$. Linearity is also to be found in rules $[\to E], [\times I], [\times EL], [\times EC]$ where linear context is split between premises of the rule. The fact that Γ is classical is achieved through the contraction rule $[CTN]$ (which has no counterpart for Δ hence insuring linearity on it) that allows the multiple use of a variable.

The premise $\cdot;\cdot \vdash M : \tau$ of rule $[AxL]$, together with the side condition $! \notin M$ is used to forbid declaration of a variable of a type τ containing a $\mathbf{Q}$ not guarded by $\Diamond$ in the classical context (thus making possible the duplication of quantum bits). It is possible to introduce $y : \Diamond\mathbf{Q}$ in the classical context because $.;. \vdash \nu x.x : \Diamond\mathbf{Q}$. The same trick is used in rule $[\times EC]$, where we check that τ and σ do not contain unguarded $\mathbf{Q}$. In this paper we have only considered two $\times$ elimination rules: one is fully classical and the other one is fully linear. It would be easy to define typing rules of couples where one member is linear whereas the other one is classical.

$[CST0]$ $\dfrac{}{\cdot;\cdot \vdash 0 : \mathbf{B}}$ $\qquad$ $[CST1]$ $\dfrac{}{\cdot;\cdot \vdash 1 : \mathbf{B}}$

$[AxC]$ $\dfrac{\cdot;\cdot \vdash M : \tau \quad !\notin M}{y : \tau;\cdot \vdash y : \tau}$ $\qquad$ $[AxL]$ $\dfrac{}{\cdot;y : \tau \vdash y : \tau}$

$[ExC]$ $\dfrac{\Gamma_1, y : \sigma', x : \sigma, \Gamma_2; \Delta \vdash M : \tau}{\Gamma_1, x : \sigma, y : \sigma', \Gamma_2; \Delta \vdash M : \tau}$ $\qquad$ $[ExL]$ $\dfrac{\Gamma; \Delta_1, y : \sigma', x : \sigma, \Delta_2 \vdash M : \tau}{\Gamma; \Delta_1, x : \sigma, y : \sigma', \Delta_2 \vdash M : \tau}$

$[WkgC]$ $\dfrac{\Gamma; \Delta \vdash M : \tau}{\Gamma, x : \sigma; \Delta \vdash M : \tau}$ $\qquad$ $[WkgL]$ $\dfrac{\Gamma; \Delta \vdash M : \tau}{\Gamma; \Delta, x : \sigma \vdash M : \tau}$

$[\to IC]$ $\dfrac{\Gamma, x : \sigma; \Delta \vdash M : \tau}{\Gamma; \Delta \vdash \lambda x.M : \sigma \to \tau}$ $\qquad$ $[\to IL]$ $\dfrac{\Gamma; \Delta, x : \sigma \vdash M : \tau}{\Gamma; \Delta \vdash \lambda x.M : \sigma \to \tau}$

$[\to E]$ $\dfrac{\Gamma; \Delta_1 \vdash M : \sigma \to \tau \quad \Gamma; \Delta_2 \vdash N : \sigma}{\Gamma; \Delta_1, \Delta_2 \vdash (M\ N) : \tau}$ $\qquad$ $[\times I]$ $\dfrac{\Gamma; \Delta_1 \vdash M : \tau \quad \Gamma; \Delta_2 \vdash N : \sigma}{\Gamma; \Delta_1, \Delta_2 \vdash \langle M, N \rangle : \tau \times \sigma}$

$[\nu I]$ $\dfrac{\Gamma; \Delta, x : \mathbf{Q} \vdash M : \sigma}{\Gamma; \Delta \vdash \nu x.M : \Diamond \tau}$ $\qquad$ $[\nu E]$ $\dfrac{\Gamma; \Delta \vdash M : \Diamond \tau}{\Gamma; \Delta \vdash !M : \tau}$

$[CTN]$ $\dfrac{\Gamma, x : \tau, y : \tau;\Delta \vdash M : \sigma}{\Gamma, z : \tau;\cdot \vdash M[x := z; y := z] : \sigma}$

$[\times EL]$ $\dfrac{\Gamma'; \Delta_1 \vdash N : \tau \times \sigma \quad \Gamma'; \Delta_2, x : \tau, y : \sigma \vdash M : \tau'}{\Gamma; \Delta_1, \Delta_2 \vdash \mathsf{let}\ \langle x, y \rangle = N\ \mathsf{in}\ M : \tau'}$

$[\times EC]$ $\dfrac{\Gamma; \Delta_1 \vdash N : \tau \times \sigma \quad \Gamma, x : \tau, y : \sigma; \Delta_2 \vdash M : \tau' \quad \begin{array}{c} \cdot;\cdot \vdash Q : \tau \\ \cdot;\cdot \vdash P : \sigma \quad !\notin \langle P, Q \rangle \end{array}}{\Gamma; \Delta_1, \Delta_2 \vdash \mathsf{let}\ \langle x, y \rangle = N\ \mathsf{in}\ M : \tau'}$

$[\rho I]$ $\dfrac{\Gamma; \Delta, \ell_1 : \mathbf{Q}, \ldots, \ell_n : \mathbf{Q} \vdash M : \tau}{\Gamma; \Delta \vdash \rho \ell_1, \ldots, \ell_n.M : \tau}$

Fig. 1. λ_{GS}^Q typing rules

Rules $[\nu I]$ and $[\nu E]$ are standard introduction and elimination rules for a connective (here $\Diamond$).

There is no elimination rule for binder ρ. This task will be fulfilled via garbage collection and since $[\rho I]$ does not modify the type of the expression, then it causes no harm to the subject reduction property.

The idea is that the programmer does not use ρ binders: they are completely managed by evaluation procedure. It causes no problem for subject reduction since there is no effect on the type of the term.

The standard subject reduction property is verified in λ_{GS}^Q.

Theorem 1 (Subject Reduction). *Let M be such that $\Gamma;\cdot \vdash M : \tau$ and $[|\varphi\rangle, \mathcal{K}, M] \rightsquigarrow [\mathcal{S}', \mathcal{K}, M']$, then $\Gamma;\cdot \vdash M' : \tau$*

Consider for instance the following piece of code :

$$cf \overset{def}{=} \nu x.\mathfrak{M}(\mathfrak{H}(x))$$

then $.;. \vdash cf : \Diamond \mathbf{B}$ is derivable. So one can use $x_{cf} : \Diamond \mathbf{B}$ in the classical typing context. Thus, it is possible to reuse x_{cf}(thanks to rule $[CTN]$). It is interesting since cf denotes a perfect coin flipping, and should be usable as many times as wanted in a probabilistic algorithm. This simple example is not possible to directly encode in [21] where qbits are linearly used: one version of x_{cf} must be defined for every use of the coin flipping or promotion must be used.

4   Conclusion

In this paper we present a variation of the pure λ-calculus: λ_{GS}. We introduce two new binders ν and ρ to bind variables that denote locations on a physical device implementing a global state over which a computation is performed. Usually variable name management (α-equivalence, fresh name generation) is handled through λ binding but it does create two kind of problems when names refer to an external device.

1. Compositionality is lost due to side effects that may occur in the external device: one cannot ensure that effects related to a given location name are limited to the lexical scope of this location name.
2. The description of an algorithm and its actual execution are two separate things in such a context. Think for instance at the no-cloning axiom in quantum computing: it only applies to *real* quantum bits, it should be possible to be able to duplicate abstract algorithms like the example of the fair coin-flipping.

Binder ρ is designed to take into account problem (1) whereas ν is designed to take into account problem (2). This distinction proves useful since it allows the definition of a type system for a quantum calculus that is more flexible than previous propositions of the literature [21,6].

This work compares to the NEW calculus of Gabbay [12] where a context substitution is introduced through a new binder. Nevertheless, this works principally addresses problems related to α-equivalence and does not talk about scope extrusion. Another related work is the short note of Baro and Maurel [10] in which the λ binder is split into two constructions: a pure binder, ν and a combinator for the implementation of the β-reduction. In this case too the scope extrusion is not addressed, and there is no notion of state.

We believe that λ_{GS} will prove useful in other situations than the one presented in this paper. As future work we plan to work on a typing system to analyze entanglement/separation properties of quantum bits in λ_{GS}^{Q}. An idea would be to decorate $\mathbf{Q}$ types with entanglement classes (two quantum bits of the same class would be supposed entangled). As entanglement is typically a non compositional property, this would give an interesting application of ρ scope extrusion property.

Another line of work is to investigate other outside computing devices than quantum ones. For instance one can try to define and study functional computing languages for DNA computers [5] or chemical computers [4].

References

1. Quantiki - introductory tutotrials:
   http://www.quantiki.org/wiki/index.php/Category:Introductory_Tutorials
2. Abadi, M., Banerjee, N., Heintze, N., Riecke, J.G.: A core calculus of dependency. In: Proceedings of the 26th Annual ACM Symposium on Principles of Programming Languages (POPL'99), pp. 147–160. ACM Press, New York (1999)
3. Abramsky, S., Honda, K., McCusker, G.: A fully abstract game semantics for general references. In: Proceedings of thirteenth Annual IEEE Symposium on Logic in Computer Science (LICS'98), pp. 334–344. IEEE Computer Society Press, Los Alamitos (1998)
4. Adamatzky, A.I.: Information-processing capabilities of chemical reaction-diffusion systems. 1. belousov-zhabotinsky media in hydrogel matrices and on solid supports. Advanced Materials for Optics and Electronics 7(5), 263–272 (1997)
5. Adleman, L.M.: Molecular computation of solutions to combinatorial problems. Science 266(11), 1021–1024 (1994)
6. Altenkirch, T., Grattage, J.: A functional quantum programming language. In: 20th Annual IEEE Symposium on Logic in Computer Science, IEEE Computer Society Press, Los Alamitos (2005)
7. Arrighi, P.: Quantum computation explained to my mother. Bulletin of the EATCS 80, 134–142 (2003)
8. Barendregt, H.P.: The Lambda Calculus; Its Syntax and Semantics. North-Holland, Revised Edition (1984)
9. Barendregt, H.P.: Lambda calculi with types. In: Abramsky, S., Gabbay, D., Maibaum, T. (eds.) Handbook of Logic in Computer Science, Clarendon Press, Oxford (1993)
10. Baro, S., Maurel, F.: The qν and qνK calculi: name capture and control. Technical Report PPS//03/11//n16, Université Paris VII (March 2003)
11. Berger, M., Honda, K., Yoshida, N.: A logical analysis of aliasing in imperative higher-order functions. In: Danvy, O., Pierce, B.C. (eds.) Proceedings of the 10th ACM SIGPLAN International Conference on Functional Programming, ICFP 2005, pp. 280–293. ACM Press, New York (2005)
12. Gabbay, M.J.: A NEW calculus of contexts. In: Proc of the 7th ACM SIGPLAN, Symposium on Principle and Practice of Declarative Programmning, PPDP'05, pp. 94–105. ACM Press, New York (2005)
13. Knill, E.: Convention for quantum pseudocode. Technical Report LAUR-96-2724, Los Alamos National Laboratory (1996)
14. Launchbury, J., Jones, S.L.P.: Lazy functional state threads. In: Proceedings of the ACM SIGPLAN'94 Conference on Programming Language Design and Implementation (PLDI'94), pp. 24–35. ACM Press, New York (1994)
15. Milner, R.: Communicating and mobile systems: the π-calculus. Cambridge University Press, Cambridge (1999)
16. Milner, R., Tofte, M., Harper, R., MacQueen, D.: The Definition of Standard ML. MIT Press, Cambridge, 1997 (revised)
17. Nielsen, M.A., Chuang, I.L.: Quantum Computation and Quantum Information. Cambridge University Press, Cambridge (2000)
18. Pitts, A.M.: Operational semantics and program equivalence. In: Barthe, G., Dybjer, P., Pinto, L., Saraiva, J. (eds.) APPSEM 2000. LNCS, vol. 2395, pp. 378–412. Springer, Heidelberg (2002)

19. Pottier, F., Simonet, V.: Information flow inference for ML. ACM Transactions on Programming Languages and Systems 25(1), 117–158 (2003)
20. Prost, F.: A static calculus of dependencies for the λ-cube. In: Proc. of IEEE 15th Ann. Symp. on Logic in Computer Science (LICS'2000), IEEE Computer Society Press, Los Alamitos (2000)
21. Selinger, P., Valiron, B.: A lambda calculus for quantum computation with classical control. In: Urzyczyn, P. (ed.) TLCA 2005. LNCS, vol. 3461, pp. 354–368. Springer, Heidelberg (2005)
22. van Tonder, A.: A lambda calculus for quantum computation. SIAM J. Comput. 33(5), 1109–1135 (2004)

Using River Formation Dynamics to Design Heuristic Algorithms*

Pablo Rabanal, Ismael Rodríguez, and Fernando Rubio

Dept. Sistemas Informticos y Computacin
Facultad de Informática
Universidad Complutense de Madrid, 28040 Madrid, Spain
prabanal@fdi.ucm.es, {isrodrig,fernando}@sip.ucm.es

Abstract. Finding the optimal solution to NP-hard problems requires at least exponential time. Thus, heuristic methods are usually applied to obtain *acceptable* solutions to this kind of problems. In this paper we propose a new type of heuristic algorithms to solve this kind of complex problems. Our algorithm is based on river formation dynamics and provides some advantages over other heuristic methods, like ant colony optimization methods. We present our basic scheme and we illustrate its usefulness applying it to a concrete example: The Traveling Salesman Problem.

Keywords: Traveling Salesman Problem, Nature-based Algorithms, Heuristic Algorithms, Ant Colony Optimization Algorithms.

1 Introduction

The nature has inspired the design of efficient computational algorithms for decades (e.g. [10,11,2,13,14,9,5,8]). Though a myriad of computational problems have been efficiently solved by conventional algorithms, nature-inspired methods are useful because they face from an alternative point of view some computational problems that are known to be especially difficult. This is the case of NP-hard problems, which are (strongly) supposed to require exponential time in the worst case to be solved. Since it is not feasible in practice to optimally solve these problems, optimal methods are substituted by *heuristic* algorithms that usually need polynomial time to provide suboptimal solutions. In this regard, we may consider the *Traveling Salesman Problem* [7,1] (TSP). This NP-complete problem has attracted a lot of attention from researchers because, on the one hand, finding optimal paths is a requirement that frequently appears in real applications and, on the other hand, it is a suitable benchmark problem to test heuristic methods where global solutions are the result of *local* interaction procedures. In particular, these methods are based on the assumption that *good*

---

* Research partially supported by the MCYT project TIN2006-15578-C02-01, the Junta de Castilla-La Mancha project PAC06-0008-6995, and the Marie Curie project MRTN-CT-2003-505121/TAROT.

hamiltonian cycles can be found by performing independent computations at each node of the graph according to some local information (e.g., by taking into account only adjacent nodes).

In this category we may find *Ant Colony Optimization* (ACO) algorithms [5,3]. In brief, they consist in copying the method used by ants to find good paths from the colony to food sources. Ants release some amount of pheromone as they move, and this pheromone attracts other ants. When an ant finds a path to the food source, it takes the food back to the colony and repeats the path afterwards, which reinforces the pheromone trail. This makes other ants follow it, which reinforces it as well. Let us note that the pheromone trail tends to be stronger in *short* paths than in long paths. This is because ants can fully traverse a short path *more times* per unit of time than a long path, which increases the amount of pheromone at each point. Moreover, let us suppose that an ant finds a path to the food source that is shorter than other paths found so far. The new path attracts some ants, which in turn reinforce the new path, and finally all ants follow the trail (that is, only the shortest path is taken). Let us note that ACO methods construct good global paths by applying a *local* scope mechanism at each location (in particular, the pheromone reinforcement).

ACO methods provide efficient heuristic algorithms to solve TSP (see [4,3]). However, they also suffer from some intrinsic problems. On the one hand, let us note that this method lies in the idea that pheromone trails interact with each other in such a way that good paths are cooperatively constructed by ants. However, this characteristic is twofold. Sometimes paths may interfere and damage each other. For instance, local cycles involving only some nodes may be created and, even worse, be reinforced by subsequent ants. On the other hand, sometimes a local oriented method like ACO may miss some simple graph peculiarities such that, if they were taken into account, then they would allow to easily find better paths. For instance, when a better path is found by ants, it takes a lot of subsequent ant movements to reinforce the new path until it is actually preferred to other older well-established paths.

Let us note that both problems are a consequence of the following property of the method: When ants decide their next movement, only the pheromone trail at each possible destination is taken into account, but the pheromone trail at the origin is essentially ignored. Let us suppose that, instead of this, the *difference* of trails between both places were considered, and this difference were required to be positive in order to choose a given edge as next step (i.e., we require that the pheromone trail is higher at the destination than at the origin). This would help to avoid both problems considered before. On the one hand, it is impossible that *all* edges in a cycle produce a positive increment, so cycles would not be formed and fed back. Let us note that most problems concerning the reciprocal interference of paths are due to local cycles. On the other hand, if the probability that an edge is taken is proportional to the increment of pheromone trail and inverse proportional to the edge cost, then a shortcut should be preferred from the exact time it is discovered by an ant: If two sequences of edges from A to B with different cost are available then the difference between the initial and final pheromone trails is obviously equal in both paths, but costs are *lower* in the shorter one. Hence, the edges in the shorter path are preferable. That is, it would not be necessary that the proportion of ants following the better path gradually increases until the pheromone trail of the shortest path is higher.

Though this alternative approach would help to overcome these problems, the following question arises: How can we make pheromone trails to be increasing along the steps of each path? In order to answer this question, it is better to give the ant metaphor up and consider another framework. Actually, the alternative framework will be based on nature as well, in particular in the *river formation dynamics*. Let us consider that a water mass is unleashed at some high point. Gravity will make it to follow a path down until it cannot go down anymore. In Geology terms, when it rains in a mountain, water tries to find its own way down to the sea. Along the way, water erodes the ground and transforms the landscape, which eventually creates a riverbed. When a strong down slope is traversed by the water, it extracts soil from the ground in the way. This soil is deposited later when the slope is lower. Rivers affect the environment by reducing (i.e. eroding) or increasing (i.e. depositing) the altitude of the ground. Let us note that if water is unleashed at all points of the landscape (e.g., it rains) then the river form tends to optimize the task of collecting all the water and take it to the sea, which does not imply to take the shortest path from a *given* origin point to the sea. Let us remark that there are *a lot of* origin points to consider (one for each point where a drop fell). In fact, a kind of *combined* grouped shortest path is created in this case. The formation of tributaries and meanders is a consequence of this. However, if water flows from a *single* point and no other water source is considered, then the water path tends to provide the most efficient way to reduce the altitude (i.e., it tends to find the shortest path).

In this paper we present an algorithm based on these ideas called RFD (from *River Formation Dynamics* algorithm) and we study how it allows to overcome the problems commented above appearing in ACO methods. Besides, we apply the method to TSP. The details of the method are discussed in the next sections. In addition, our approach is compared with ACO and some results are reported. It is worth to point out that, though the development of RFD is still in a preliminary phase and has not been fully optimized yet (only some improvements, out of a big set of available choices, have been implemented), current results are especially promising. In particular, we observe the following property: Though the time required to converge to a solution is a bit longer in RFD than in ACO, the quality of solutions provided by RFD is better in general.

The rest of the paper is structured as follows. In the next section we present the general algorithm developing these ideas. Then, in Section 3 this method is particularized for TSP. Afterwards, in Section 4 we present an implementation of the algorithm and we compare the results of our method with those obtained using the ACO method. Finally, in Section 5 we present our conclusions and some lines of future work.

2   General Method

In this section we present our algorithm inspired on river formation dynamics. It finds short paths in graphs, in general, and good solutions for TSP, in particular. The particularization of the general scheme to TSP will be discussed in the next section.

The method works as follows. Instead of associating pheromone values to edges, we associate *altitude* values to nodes. Drops erode the ground (they reduce the altitude of nodes) or deposit the sediment (increase it) as they move. The probability of the drop to take a given edge instead of others is proportional to the gradient of the down slope in the edge, which in turn depends on the difference of altitudes between both nodes and the distance (i.e. the *cost* of the edge). At the beginning, a flat environment is provided, that is, all nodes have the same altitude. The exception is the destination node, which is a *hole*. Drops are unleashed at the origin node, which spread around the flat environment until some of them fall in the destination node. This erodes adjacent nodes, which creates new down slopes, and in this way the erosion process is propagated. New drops are inserted in the origin node to transform paths and reinforce the erosion of promising paths. After some steps, good paths from the origin to the destination are found. These paths are given in the form of sequences of decreasing edges from the origin to the destination.

This method provides the following advantages with respect to ACO. On the one hand, local cycles are not created and reinforced because they would imply an *ever decreasing cycle*, which is contradictory. Though ants usually take into account their past path to avoid repeating nodes, they cannot avoid to be led by pheromone trails through some edges in such a way that nodes may be repeated in the next step[1]. On the contrary, *altitudes* cannot lead drops to these situations. On the other hand, when a shorter path is found in RFD, the subsequent reinforcement of the path is fast: Since the same origin and destination are concerned in both the old and the new path, the difference of altitude is the same but the distance is different. Hence, the edges of the shorter path necessarily have higher down slopes and are immediately preferred (in average) by subsequent drops. Moreover, let us note that the erosion process provides a method to avoid inefficient solutions: If a path leads to a node that is lower than any adjacent node (i.e., it is a blind alley) then the drop will deposit its sediment, which will increase the altitude of the node. Eventually, the altitude of this node will match the altitude of its neighbors, which will avoid that other drops fall in this node. Moreover, more drops could be cumulated in the node until the mass of water reaches adjacent nodes (a *lake* is formed). If water reaches this level, other drops will be allowed to cross this node from one adjacent node to another. Thus, paths will not be interrupted until the sediment fills the hole. This provides an implicit method to avoid inefficient behaviors of drops.

---

[1] Usually, this implies either to repeat a node or to *kill* the ant. In both cases, the last movements of the ant were useless.

2.1   Basic Algorithm

The basic scheme of the algorithm follows:

```
initializeDrops()
initializeNodes()
while (not allDropsFollowTheSamePath()) and (not otherEndingCondition())
    moveDrops()
    erodePaths()
    depositSediments()
    analyzePaths()
end while
```

The scheme shows the main ideas of the proposed algorithm. We comment on the behavior of each step. First, drops are initialized (`initializeDrops()`), that is, all drops are put in the initial node. Next, all nodes of the graph are initialized (`initializeNodes()`). This consists of two operations. On the one hand, the altitude of the destination node is fixed to 0. In terms of the river formation dynamics analogy, this node represents the *sea*, that is, the final goal of all drops. On the other hand, the altitude of the remaining nodes is set to some equal value.

The `while` loop of the algorithm is executed until either all drops find the same solution (`allDropsFollowTheSamePath()`), that is, all drops traverse the same sequence of nodes, or another alternative finishing condition is satisfied (`otherEndingCondition()`). This condition may be used, for example, for limiting the number of iterations or the execution time. Another choice is to finish the loop if the best solution found so far is not surpassed during the last n iterations.

The first step of the loop body consists in moving the drops across the nodes of the graph (`moveDrops()`) in a partially random way. The following *transition rule* defines the probability that a drop k at a node i chooses the node j to move next:

$$P_k(i,j) = \begin{cases} \frac{decreasingGradient(i,j)}{\sum_{l \in V_k(i)} decreasingGradient(i,l)} & \text{if} \quad j \in V_k(i) \\ 0 & \text{if} \quad j \notin V_k(i) \end{cases} \tag{1}$$

where $V_k(i)$ is the set of nodes that are *neighbors* of node i that can be visited by the drop k and *decreasingGradient(i,j)* represents the negative gradient between nodes i and j, which is defined as follows:

$$decreasingGradient(i,j) = \frac{altitude(i) - altitude(j)}{distance(i,j)} \tag{2}$$

where *altitude(x)* is the altitude of the node x and *distance(i,j)* is the length of the edge connecting node i and node j. Let us note that, at the beginning of the algorithm, the altitude of all nodes is the same, so $\sum_{l \in V_k(i)} decreasingGradient(i,l)$ is 0. In order to give a special treatment to flat gradients, we modify this scheme as follows: We consider that the probability that a drop moves through an edge with 0 gradient is set to some (non null) value. This enables drops to spread around a flat environment, which is mandatory, in particular, at the beginning of the algorithm.

In the next phase (`erodePaths()`) paths are eroded according to the movements of drops in the previous phase. In particular, if a drop moves from node A to node B then we erode A. That is, the altitude of this node is reduced depending on the current gradient between A and B. In particular, the erosion is higher if the down slope between A and B is high. If the edge is flat then a small erosion is performed. The altitude of the final node (i.e., the *sea*) is never modified and it remains equal to 0 during all the execution.

As we commented before, the erosion process avoids in practice that drops follow cycles because a cycle must include at least one increasing slope and, according to the basic behavior presented before, drops cannot climb them up. However, in ACO pheromone trails involving independent paths can interfere with each other in such a way that ants feed cycles back and follow them[2].

In addition, let us suppose that a new path, which is better than other paths considered so far, is found. In RFD, the erosion quickly favors it against other choices. This is because down gradients in the new path are preferable *in average* to those of old paths: Globally, a different cumulated distance is traversed in both paths, but the difference of altitudes is the same. This eases the subsequent task of feeding the new path back so that not only steps are preferred in average, but also each of them is individually preferred to other older choices. On the contrary, when an ant finds a better path for the first time, the pheromone trail may still be negligible compared to other older paths. Hence, this path must be gradually fed back until the steps of the new path are preferred at least *on average*.

Once the erosion process finishes, the altitude of all nodes of the graph is slightly increased (`depositSediments()`). The objective is to avoid that, after some iterations, the erosion process leads to a situation where all altitudes are close to 0, which would make gradients negligible and would ruin all formed paths.

Finally, the last step (`analyzePaths()`) studies all solutions found by drops and stores the best solution found so far.

2.2   Basic Improvements

In this section we show some improvements we apply to the previous basic scheme. As we said before, we allow drops to move through edges with a gradient equal to 0. Going one step further, we apply the following improvement: We let drops climb *increasing* slopes with a low probability. This probability will be inverse proportional to the increasing gradient. This new feature improves the search of good paths. Let us note that solutions found during the first steps

---

[2] Let us note that, in order to feed a cycle back, it is not necessary that an ant actually follows the cycle. It is enough that each ant reinforces a *different* part of the cycle in such a way that all parts are individually reinforced.

tend to bias the exploration of the graph afterwards. This is because previously formed paths tend to be followed by subsequent drops. Enabling drops to climb increasing slopes with some low probability allows to find alternative choices and enables the exploration of other paths in the graph. This partially decouples the method from its behavior in the first steps. Actually, the probability of climbing increasing slopes encapsulates most of the dependency of the method on previous solutions in a single value. As usual in heuristic search algorithms, this dependency must provide a suitable tradeoff between past solutions and alternative choices.

In addition, the probability of climbing increasing slopes will be reduced during the execution of the algorithm, that is, for each new iteration this probability is slightly reduced. After performing some iterations, the exploration of the graph will be sufficient, so the necessity of searching more alternative choices will be reduced (recall that we reduce the probability of climbing slopes, but the probability of following edges which do *not* have the highest decreasing gradient remains the same). Besides, we borrow the following idea from *Simulated Annealing* methods (see [10,6]). In addition to constantly reducing the probability that a drop takes an increasing slope, we add the following feature: Every N iterations, this probability is increased instead of slightly reduced. On the one hand, this reduces the dependency on (bad) older well established solutions because we periodically enable to search other alternative choices. On the other hand, paths that are actually good will survive to this periodic *tilting* action. Let us note that bad paths in the short term may turn out to be good in the long term. So, this technique helps to avoid getting blocked in local optima. The probability of taking increasing slopes is incremented in such a way that decrements surpass them in the overall, that is, this probability globally tends to 0.

Enabling drops to climb slopes implies that the probability that a drop traverses a local cycle is not 0. However, the method still provides an effective way to avoid that drops traverse cycles in practice. On the one hand, the probability of this is low because a cycle must contain at least one increasing slope. In particular, if there is only one increasing slope then it must compensate the rest of decreasing slopes, so its increasing gradient must be high. Since the probability of climbing an increasing slope up is inverse proportional to the gradient, the probability of climbing this slope is very low. Moreover, as we said before, the probability of climbing up slopes is reduced along time. On the other hand, the probability that an ant follows a local cycle in ACO *increases* as long as all edges of the cycle are still being reinforced (recall that each edge could be reinforced by an ant following a different path).

We also enable drops to *deposit* sediment in nodes, which was not considered in the basic scheme presented before. This happens when all movements available for a drop imply to climb an increasing slope and the drop fails to climb any edge (according to the probability assigned to it). In this case, the drop is blocked and it deposits the sediments it transports. This increases the altitude of the current node. The increment is proportional to the amount of cumulated sediment. As we said before, this allows to gradually *punish* paths leading to blind alleys.

Eventually, the cumulated sediment increases the altitude of this node until it matches the altitude of nearby nodes. At this moment, slopes leading to this node will not be decreasing anymore. This will prevent other drops from following this path.

The last improvement consists in *gathering* drops to reduce the number of individual movements. Drops in the same node are gathered into a single drop with higher size (i.e., the *caudal* is increased). In this way, we can handle a single drop of size N instead of N drops of size 1. At the beginning, a drop of size N is located in the the first node. Later, when the drop is moved and there are several choices to move next, the drop is split into drops of smaller size, each of them following a different path. Gathering drops allows to reduce the number of decisions made by the algorithm at each step because there are less drops to consider: Instead of spending computational effort in moving each drop, a single decision for each group is taken. This improves the efficiency of the method. Let us note that, in the worst case, a drop is split into drops of size 1 and each of them does not join an alternative caudal afterwards, that is, the size of each remains equal to 1. Thus, we have the case where drops are not gathered.

When we move a big drop we act as follows. Let us assume that $\texttt{trunc}(X)$ returns the integer part of X. First, we consider the probabilities of taking each choice if the size of the drop were 1. Then, we use these probabilities to *deterministically* split the drop into parts, moving each of them to each destination. This is done as follows: If a drop of size 204 is at node A and its available destinations are B, C, and D, with probabilities 0.35, 0.22, and 0.43, respectively, then a drop of size $\texttt{trunc}(204 \cdot 0.35) = \texttt{trunc}(71.40) = 71$ moves to A, a drop of size $\texttt{trunc}(204 \cdot 0.22) = \texttt{trunc}(44.88) = 44$ moves to C, and a drop of size $\texttt{trunc}(204 \cdot 0.43) = \texttt{trunc}(87.72) = 87$ moves to D. The remaining $204 - (71 + 44 + 87) = 2$ drops are (randomly) moved as single drops by applying formula (2) presented in the previous section.

3   Applying the Method to TSP

All previous discussions concern the application of the method to the general case, that is, to find a good path in a graph from a node to another. However, in this section we will apply this approach to TSP. We can summarize TSP as follows: Given a set of cities and the costs of traveling from any city to any other city, compute the cheapest round-trip route that visits each city exactly once and then returns to the starting city. Note that, since we are looking for a cyclic tour, it is irrelevant what concrete node is the origin of the tour: The cycle A-B-C-D-A has the same length as B-C-D-A-B or C-D-A-B-C.

Facing this problem requires to particularize the previous general framework. First, let us remark that the goal of TSP consists in finding a *cycle* in the graph: We traverse all the nodes (cities) of the graph and return to the departure node. However, in RFD any cycle must include at least one increasing slope. Since solutions are given in the form of sequences of decreasing edges, the basic scheme presented in the previous section must be slightly adjusted. In order to allow a decreasing slope from a node to itself, we will *clone* the origin/destination node into two separated nodes. In this way, each instance will be able to have a different altitude. In fact, the cloned node will have altitude 0, which represents that it is the destination node.

A second special feature of the problem is that we need to find a path visiting all the nodes of the graph. Thus, each drop needs memory to remember what nodes have not been visited yet. Let us remark that the memory is also needed when using ACO because ants also need to remember what nodes have not been visited yet.

In the general scheme presented in the previous section, nodes are eroded immediately after moving each drop. However, in the adaptation to TSP presented in this section drops will not erode nodes immediately after each movement. On the contrary, the erosion of the path will not be performed until the drop reaches the final destination. At this time, all nodes traversed in the path will be eroded according to the gradients of the corresponding edges. Let us suppose that a solution A-B-C-D was found by a drop. Then, the altitude of A is reduced according to the gradient between A and B, the altitude of B is reduced according to the gradient between B and C, and so on. The path traversed by the drop is extracted from its memory. Note that some drops will not find a solution in their tours. In that case, the drop *will be evaporated* or *it will produce sediments*, depending on the reason why the drop did not find a complete path. Thus, for each drop there are three possibilities:

(a) The drop *finds a solution*. In this case the drop finds a way to traverse all the cities passing through each city only once and returning to the city of origin. Thus, the drop actually erodes the path it has followed.
(b) The drop *evaporates*. This is done when the drop does not find a solution because, at a given moment, all adjacent cities had already been visited. In this case, the drop does not contribute with information to the solution, that is, the path it traversed is not eroded.
(c) The drop *deposits sediments*. This is done when the drop does not find a solution because, at a given moment, the altitude of all the adjacent cities is higher than the altitude of the current city. In this case, the drop deposits its sediments in this *valley* city, increasing its height.

An additional improvement can be considered when eroding a solution in case (a): Nodes can be eroded in proportion to the *quality* of the solution found by the drop (that is, in inverse proportion to the overall length of the path). In this way, decreasing slopes are eroded faster in good paths. In particular, the erosion of a bad edge (e.g. with a low decreasing slope) is high if the edge is actually part of a good path.

Finally, let us comment on an important detail needed to solve TSP by using RFD. Let us suppose that a solution A-B-C-D-E-A' is found (being A' the clone of A playing the role of *destination* node). Drops create a decreasing gradient along this path. In particular, the altitude of B is higher than the altitude of C, this one is higher than D, and D is higher than E. Let us also suppose that

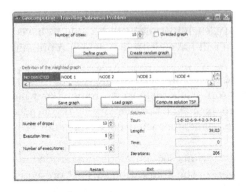

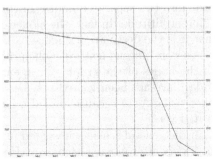

Fig. 1. Application interface   **Fig. 2.** Altitudes in a solution

there exists an edge from B to E. Since the difference of altitude between B and E is the addition of the differences between B and C, C and D, and D and E, the decreasing gradient from B to E could be so big that drops *prefer* to go directly from B to E. However, in that case drops will fail in finding a solution: C, D, and E would not be included in this path, but solutions must traverse all nodes. In order to avoid that the altitude of adjacent nodes wrongly deviates formed paths, an auxiliary node will be created at each edge of the graph. These auxiliary nodes, that we call *barrier nodes*, will also be part of the solution. In fact, barrier nodes being part of a solution will be eroded (like the rest of nodes), while barrier nodes that are not part of a solution will not be eroded. In the previous example, the solution A-B-C-D-E-A' will be represented as A-ab-B-bc-C-cd-D-de-E-ea'-A', being ab the barrier node appearing between A and B, and so on. These barrier nodes will be eroded when finding this solution. However, the barrier node between B and E, be, will not be eroded by drops following this path. Drops directly moving from B to E do not find a solution, so node be will not be eroded by an alternative path. Thus, the altitude of node be will remain high and drops at B will not prefer moving to be. That is, be imposes a *barrier* between B and E.

Let us note that barrier nodes must be taken into account in the initialization and sedimentation phases of the algorithm. In the initialization phase, we must set the height of barrier nodes. This height will be the same as the height of the rest of nodes of the graph. Regarding the sedimentation phase, it will be necessary to increase their heights in the same way as any other node.

4   TSP Implementation and Results

In this section we present an implementation of our approach to solve TSP and we compare the experimental results of our algorithm and ACO. We identify the situations where RFD surpasses ACO and we study the reasons for this. All the experiments were performed in an Intel T2400 processor with 1.83 Ghz.

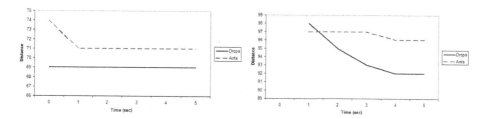

Fig. 3. Results for a 20 (left) and 30 (right) nodes random graph

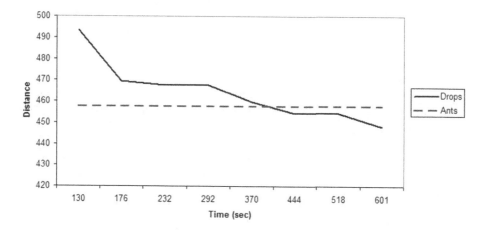

Fig. 4. Results for the TSPLIB ei151 graph

Figure 1 shows the interface of the application we developed to implement both methods and Figure 2 depicts an example of solution found by RFD. It shows the altitudes of the nodes of the path in the order they are traversed.

Let us present our experimental results. Figure 3 (left) shows the results of an experiment where the input of both algorithms was a given randomly generated graph with 20 nodes. The graphic shows the best solution found by each algorithm for each execution time. Figure 3(right) shows the results obtained after applying both methods to a random graph with 30 nodes. Next, in Figure 4 we show the results obtained for a graph taken from the TSPLIB library [12]. TSPLIB is a library of sample instances for TSP from various sources and of various types. Nodes are defined by means of points in a 2D plain and there is an edge between all pairs of nodes. The distance of each edge is the Euclidean distance between both points of the plain. Figure 4 shows the results after applying the graph ei151 from TSPLIB (with 51 nodes) to both methods. The basic shape shown in the figure is usually obtained with more benchmark examples.

Before we analyze the results shown in these figures, let us comment on how the ACO approach and the RFD approach were compared in each experiment. The number of ants in ACO and the number of drops in our approach does not coincide in each experiment. For each graph, we analyzed the number of ants (respectively drops) that provided the *best* solutions in the ACO algorithm (resp. in RFD). We observed that the *optimal* number of entities traversing each graph was different in both approaches. For example, for the graph with 51 nodes, the number of ants was set to one thousand and the number of drops was set to one hundred because we observed that one thousand and one hundred were the optimal choices for each method. Let us note that moving an ant does not require exactly the same computational effort than moving a drop. Moreover, the effect caused by moving each of them in terms of its partial contribution to future solutions (i.e., in terms of higher pheromone trails and lower altitudes, respectively) is different. Since both entities are not comparable, assessing the performance of both algorithms with the same number of populations is not appropriate. On the contrary, we compare both methods by applying the optimal population in each case.

We extract the following conclusions from the experimental results. We observe that ACO provides good solutions in short times. This is because ants tend to converge through the same path in short times. On the other hand, the solutions found by RFD after short execution times are poor. However, after some time, the quality of the solutions of the RFD method surpasses the quality of the solutions provided by ACO. That is, the convergence of drops is achieved later than the convergence of ants, but better paths are traversed by drops. Hence, RFD is the best choice if the quality of the solutions is more important than the time needed to achieve them.

There are several reasons for this difference in both approaches. First, let us note that slopes formed by drops tend to form a general *movement tendency* for subsequent drops. This tendency leads drops from the origin node to itself (in our method, to its *clone*) in an established direction. It is similar to the case that we take a Billiard table and we gradually tilt it towards a corner pocket. This tendency allows drops to homogeneously explore the graph after some iterations, that is, edges are explored regardless of their distance to the origin/destination node. However, this general tendency is lower in ACO methods because pheromone trails in edges do not impose a given direction to ants. If an edge from A to B has a high pheromone trail then ants at A try to go to B, and ants at B tend to go to A. Hence, general tendencies to explore the graph are more difficult to create, and ants tend to cumulate themselves in specific local areas because the exploration of each zone feeds itself back. This allows to improve the exploration of areas already traversed, but makes it more difficult to explore other alternative zones. The result is that the convergence is faster in the ACO method than in RFD, but the solutions found by ACO are worse because the exploration is focalized.

Another reason for this behavior is related to the way a new shortcut is reinforced in our method. As we already commented in previous sections, when a shorter path is found by drops, it is easy for subsequent drops to reinforce this path until it is preferred in all cases to other choices. This is because the new path is actually attractive for other drops from the time the reinforcement begins. We could argue that, if shorter paths are reinforced quickly, then the preference of drops for short paths would *accelerate* their convergence on the cost of ignoring other paths that are bad in the short term (but may be good in the long term). That is, we have a reason to obtain an *opposite* behavior to the one commented before. However, we observed the following behavior: Since short paths quickly become attractive choices, it is usual that, at the same time, there are *several* new attractive choices challenging older longer paths. As long as these new alternative paths confront each other, drops have several attractive choices to move next. This enables a deeper exploration of the graph (and, paradoxically, a slower overall convergence).

The way in which we handle the sedimentation in RFD also favors this behavior. Recall that we allow drops to deposit sediments in blind alleys. Eventually this *fills* the hole, which avoids that other drops follow the same wrong path. As a consequence, drops perform a *try and fail* mechanism where new paths are iteratively formed/blocked until good paths are found. This also enables a deeper exploration of the graph, leading to better solutions and longer execution times.

As we commented in previous sections, our method avoids in practice that drops follow local cycles, which would lead to situations where all edges available from a node connect to repeated nodes. Being provided with an intrinsic method to avoid the repetition of nodes (in our case, the difference of altitudes) lets drops spread around the graph, which also improves the exhaustiveness of the exploration. However, the main consequence of this characteristic is the following: The number of drops that must be eliminated because they lead to repeated nodes is *lower* than the number of ants that must be eliminated for the same reason.

Let us note that in RFD drops are also eliminated when all adjacent nodes are higher than the current node. Thus, there is a reason to eliminate drops in RFD that does not exist in ACO. However let us note that, when a drop is blocked in a valley, the drop deposits its sediment. That is, the drop contributes to *transform* the graph values and its existence is actually useful for subsequent drops. On the other hand, a blocked ant is useless for the method because it disappears without modifying any pheromone trail (note that an ant modifies trails *only* when it finds a solution). Hence, the computational effort spent to move it was lost.

5   Conclusions and Future Work

In this paper we have presented a new heuristic method to solve complex problems. The method is inspired on the way rivers are created in nature. By using it, we can obtain acceptable solutions to NP-hard problems in reasonable times. Our method obtains competitive results when compared with other more matured schemes, like ants colonies. In the TSP case study, our current experimental results shows that the RFD method is preferable to ACO if the optimality of the solution is the main requirement. The fast reinforcement of shortcuts, the avoidance of local cycles, the punishment of wrong paths, and the creation of global direction tendencies motivate these results.

Our current lines of work are divided in two parts. First, we are working on optimizing our basic scheme. For instance, we may consider the following new directions of improvement: Masses of water could be considered as part of the altitude of nodes (currently, only the altitude of the ground is considered); the number of drops moved across the environment, which currently is constant from the first step, could vary along the execution depending on the convenience of parallel/focused exploration; in order to improve the mechanism to avoid wrong paths, sediments could be deposited not only in standard nodes but also in barrier nodes; an hybrid ants/drops approach, where the weight of each method varies along the execution, could be constructed. This would allow to get the best characteristics of each method in a single approach.

Second, we are also applying RFD to solve a wider set of problems. First, we are considering the exploration of *dynamic* graphs, that is, graphs where edges may appear and disappear along time. We have reasons to expect that this will be a suitable application case for our method. If an old path is suddenly blocked then the sedimentation mechanism and the way in which shortcuts are reinforced will allow drops to quickly avoid the blocked path and find alternatives. In addition, we want to explore the problem of finding optimal *grouped* solutions from several origins to a single destination. As we commented in the introduction, rivers do not follow optimal paths from a single point to the sea. On the contrary, they optimize the task of gathering and transporting the water from each point where it rains (i.e., millions of departure points) to the sea through a *grouped* path. Following this idea, we can apply RFD to the problem of finding optimal paths for *public transportation lines*, that is, to decide the best path to gather and transport population from residential areas to e.g. the city center.

Acknowledgments

We would like to thank the anonymous reviewers for valuable comments on a previous version of this paper. Moreover, we would also like to thank Manuel Núñez and Natalia López for interesting discussions on the topic of the paper.

References

1. Applegate, D.L., Bixby, R.E., Chvatal, V., Cook, W.J.: The Traveling Salesman Problem: A Computational Study. Princeton University Press, Princeton, NJ (2006)
2. Davis, L. (ed.): Handbook of genetic algorithms. Van Nostrand Reinhold, New York (1991)
3. Dorigo, M.: Ant Colony Optimization. MIT Press, Cambridge (2004)
4. Dorigo, M., Gambardella, L.M.: Ant colonies for the traveling salesman problem. BioSystems 43(2), 73–81 (1997)
5. Dorigo, M., Maniezzo, V., Colorni, A.: Ant system: optimization by a colony of cooperating agents. IEEE Transactions on Systems, Man and Cybernetics, Part B 26(1), 29–41 (1996)

6. Fleischer, M.: Simulated annealing: past, present, and future. In: Proceedings of the 27th conference on Winter simulation, pp. 155–161 (1995)
7. Gutin, G., Punnen, A.P.: The Traveling Salesman Problem and Its Variations. Kluwer Academic Publishers, Dordrecht (2002)
8. Haupt, R.L., Haupt, S.E.: Practical Genetic Algorithms. Wiley-Interscience, New York, NY, USA (2004)
9. Kennedy, J., Eberhart, R.: Particle swarm optimization. In: Proceedings of IEEE International Conference on Neural Networks, 1995, vol. 4 (1995)
10. Kirkpatrick, S., Gelatt, Jr., C.D., Vecchi, M.P.: Optimization by Simulated Annealing. Science 220(4598), 671 (1983)
11. Langton, C.: Studying artificial life with cellular automata. Physica D 22, 120–149 (1986)
12. Reinelt, G.: TSPLIB 95. Technical report, Research Report, Institut für Angewandte Mathematik, Universität Heidelberg, Heidelberg, Germany (1995), http://www.iwr.uni-heidelberg.de/groups/comopt/software/TSPLIB95/
13. Rosenberg, G., Salomaa, A. (eds.): Lindenmayer Systems: Impacts on Theoretical Computer Science, Computer Graphics, and Developmental Biology. Springer, Heidelberg (1992)
14. Wolfram, S.: Cellular Automata and Complexity. Addison-Wesley, London, UK (1994)

Principles of Stochastic Local Search

Uwe Schöning

Institute of Theoretical Computer Science
Ulm University
89069 Ulm, Germany
uwe.schoening@uni-ulm.de

Abstract. We set up a general generic framework for local search algorithms. Then we show in this generic setting how heuristic, problem-specific information can be used to improve the success probability of local search by focussing the search process on specific neighbor states. Our main contribution is a result which states that stochastic local search using restarts has a provable complexity advantage compared to deterministic local search. An important side aspect is the insight that restarting (starting the search process all over, not using any information computed before) is a useful concept which was mostly ignored before.

Keywords: local search, stochastic local search, heuristic information, restart, meta-algorithmics.

1 Introduction

As compared to standard algorithm design techniques, local search algorithms perform a quite unconventional computation since in most cases the problem-specific evaluation function (for example, the number of unsatisfied constraints, the length of the TSP tour found so far, etc.) can only give a weak guidance information regarding what is the best direction for choosing the next local search step. Therefore, especially to escape the situation of getting stuck in local optima, up to a certain amount, randomness is used to give the search process the chance to find a way out of "dead ends." But randomness alone (so-called "uninformed search") does not help either, and is a poor strategy.

In this work we set up a general framework for modelling local search algorithms (which is more or less the standard one.) Then, under the assumption, that problem-specific ("heuristic") information about the problem in question is available which allows to focus the search on a certain well-chosen subset of the neighbors of a state, a rigorous probabilistic analysis of local search is possible. The master example where such heuristic information indeed is available, is SAT, or k-SAT. But also for other types of problems this approach might be applicable. Our analysis results in precise probability estimations for finding a solution, and shows that stochastic local search has a provable complexity advantage as

S.G. Akl et al.(Eds.): UC 2007, LNCS 4618, pp. 178–187, 2007.
© Springer-Verlag Berlin Heidelberg 2007

compared to deterministic (backtracking-like) local search. An additional outcome of the analysis is that it is much better to restart the search process several times, just after a few unsuccessful local search steps, than to continue the local search process towards the stationary (Markov chain) distribution. This is in contrast to many text book opinions where one can find the statement that restarting a local search process is usually a "poor strategy."

2   Modelling Local Search Algorithms

The following first definitions are standard when modelling the scenario of local search (see e.g. [1,8,9,10,13].)

Definition. A *local search scenario* for a combinatorial decision (or optimization problem) Π is given by the following components. Given an input instance π, one can associate with π the following items

- a *search space* S. The elements of S are called *candidate solutions* or *states*,
- a set $S_0 \subseteq S$ of *(optimal) solutions*,
- a *neighborhood relation* $N \subseteq S \times S$. (We sometimes use N in unary (functional) form so that $N(x)$ denotes the set of all y such that $(x, y) \in N$.)
- the *solution distance* $d(x)$ of a state $x \in S$ is defined by

$$d(x) = \min \{k \mid \exists x_1, \ldots, x_k \in S :$$
$$(x, x_1) \in N, (x_1, x_2) \in N, \ldots, (x_{k-1}, x_k) \in N, x_k \in S_0\}$$

Example. The well known problem SAT is, given a formula π in propositional logic, find an assignment a for the Boolean variables occuring in π which satisfies formula π. Let the formula π have n Boolean variables. Then the candidate solutions are all the possible assignments to these variables. As it is often the case, the candidate solutions, the elements of the search space S, can be divided into n components $a = (a_1, a_2, \ldots, a_n)$, $a_i \in \{0, 1\}$. For each component there are, in this case, two possible selections, namely, 0 and 1. (For more general *constraint satisfaction problems* there might be more selections per component.) Therefore, in this case, the search space is $S = \{0, 1\}^n$, and $|S| = 2^n$. The solution set is $S_0 = \{a \in S \mid a$ satisfies $\pi\}$. Two assignments a, a' are neighbors of each other, i.e. $(a, a') \in N$, and $(a', a) \in N$, if and only if they differ in exactly one component. In other words, if their Hamming distance is 1. The solution distance of an assignment a is the minimal number of bits that have to be flipped to arrive at a satisfying assignment. This is the smallest possible number of local search steps, starting from x, that might lead to a solution.

The above definitions are specific to the respective problem Π (although for the same problem Π there might be several useful definitions of the state space and the neighborhood relation.) More towards the design of a particular local search method for solving Π are the following definitions (and therefore we have separated them from the above definitions.) We already define a basic stochastic local search algorithm for Π if we specify the following items, possibly using problem-specific, heuristic information:

- An *initial probability distribution* D on S. (Let $P_D(x)$ denote the probability that candidate $x \in S$ is selected under this initial distribution.)
- for each $x \in S$, a *local search probability distribution* $L(x)$ on the set of neighbors of x, i.e. on $N(x) = \{y \in S \mid (x,y) \in N\}$. (Let $P_{L(x)}(y)$ denote the probability that y is selected in the local search step starting with the state x, where $P_{L(x)}(y) = 0$ if $(x,y) \notin N$.) We allow the situation $P_{L(x)}(y) = 0$ for all $y \in N(x)$, or also the case $N(x) = \emptyset$, then we let $P_{L(x)}(\perp) = 1$ where $\perp$ is a special state which indicates "failure".

Once these probability distributions are specified, the generic local search algorithm for problem Π works as follows.

INPUT π
Choose an initial candidate solution $x \in S$ according to distribution D
WHILE ($x \notin S_0$ AND $x \neq \perp$) DO
    Replace x by one of its neighbors $y, (x,y) \in N$, selected according
    to distribution $L(x)$
END-WHILE

In the case of an optimization problem Π it is not possible to test whether x is an optimal solution, i.e. $x \in S_0$, efficiently. In this case often an evaluation function f is used which evaluates the quality of the obtained candidate x. The value of f becomes optimal (maximal or minimal) iff $x \in S_0$. If the value $f(x)$ did not change for some time, this might be a criteria for stopping. Also, the definition of $L(x)$ might be based on the f-values of x and its neighbors $y \in N(x)$. In the simplest implementations it might be the case that $P_{L(x)}(\perp) = 1$ if there is no neighbor $y \in N(x)$ with better f-value than $f(x)$. This describes the well known problem of getting stuck in some local optimum, an issue which is the main concern in several books and publications on local search [1,3,8,9].

The following illustrates this problem from a different perspective. Assume, starting from initial candidate x_0, we follow some path $x_0, x_1, \ldots, x_k$ such that $(x_0, x_1) \in N, (x_1, x_2) \in N, \ldots, (x_{k-1}, x_k) \in N$, and $d(x_0) = k, d(x_1) = k - 1, \ldots, d(x_k) = 0$, i.e. $x_k \in S_0$. At the same time we observe the values $f(x_0), f(x_1), \ldots, f(x_k)$. When we normalize both sequences (the $d(.)$-values, and the $f(.)$-values) such that both start with 1 and end with 0, the picture might look like this:

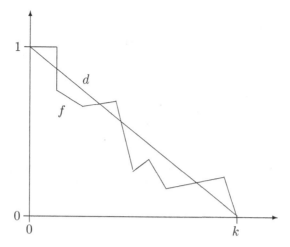

This means, except from the starting and ending point and a general tendency there are only few similarites between both curves. On the other hand, to compute the function $d(x)$ exactly is just as hard as solving the original problem Π. If we could calculate $d(y)$ exactly for all the neighbors of x we can select that one which minimizes the solution distance.

Example. In the SAT example mentioned above, a natural (and often used) way to define $f(x)$ for an assignment x, and a given formula π in conjunctive normal form (clause form) is to let

$$f(x) \;=\; \text{number of unsatisfied clauses in } \pi \text{ under assignment } x$$

Indeed, as the above figure indicates, there is not very strong correlation between $f(x)$ and the solution distance $d(x)$.

A direct, naive way of defining the inititial distribution for SAT is to assign each assignment $x \in \{0,1\}^n$ the same probability, $P_D(x) = 2^{-n}$. A similar direct way of defining the local search probability is to assign each neighbor of x the same probability, i.e.

$$P_{L(x)}(y) \;=\; \begin{cases} 1/n, & \text{if } y \in N(x), \\ 0, & \text{otherwise} \end{cases}$$

This does not result in a good local search algorithm. The probability that such a random local search step according to distribution $L(x)$ will decrease the solution distance is $d(x)/n$, that is, the closer x is to the desired solution, the harder is it to come even closer.

3   Heuristic Information and Markov Chain Analysis

In [11,15,16], in the case of k-SAT, a different local search probability distribution is used. This restricted neighborhood can be understood as bringing in some problem-specific heuristic information. Given an assignment x which does not yet satisfy the k-CNF formula π, there will be at least one clause in π which is not satisfied. Let σ be one (e.g. the first) such clause. The clause σ contains k variables (negated or non-negated). Let $N_\sigma(x)$ be the subset of $N(x)$ of size k which contains those neighbors of x where exactly one bit of x is flipped and this bit belongs to a variable in σ. We let

$$P_{L(x)}(y) = \begin{cases} 1/k, & \text{if } y \in N_\sigma(x), \\ 0, & \text{otherwise} \end{cases}$$

In this case, using $L(x)$ for a local search step, the probability of decreasing $d(x)$ by 1 is (at least) $1/k$ (since in a satisfying assignment there should at least one of the k bits be flipped to satisfy the clause σ), and this is better than the above naive approach once the solution distance $d(x)$ is at most n/k. Indeed, as shown in [11], the expected number of local search steps until a solution is found, is quadratic in n, in the case of 2-SAT. The case $k > 2$ is handled in [15,16]. The probability that a single run of $3n$ local search steps will find a solution is shown to be at least $\frac{1}{p(n)} \cdot (\frac{k}{2k-2})^n$ where p is some (small) polynomial.

It is clear that this basis version of a local search algorithm will only be successful with some probability, depending on the choice of D and $L(x)$, as indicated by the SAT example above. Therefore, it is a good idea to restart the basis algorithm several times with independent new initial solution candidates. This is realized in the following program (here t is some parameter which indicates the number of local search steps until to restart.)

```
INPUT π
REPEAT FOREVER
    Choose an initial candidate solution x ∈ S according to
    distribution D; set i := 0
    WHILE ( x ∉ S₀ AND x ≠⊥ AND i ≤ t) DO
        i := i + 1
        Replace x by one of its neighbors y, (x, y) ∈ N,
        selected according to distribution L(x)
    END-WHILE
    IF x ∈ S₀ THEN STOP successfully
REPEAT-END
```

The behavior of this stochastic local search algorithm can be understood as a Markov chain in the following way.

We can understand the probability distribution D as a vector of length $|S|$. By $[D]_x$ we indicate the component indexed by $x \in S$ in D, i.e. $[D]_x = P_D(x)$.

Similarly, we can understand the set of all $L(x)$ for $x \in S$ as a $|S| \times |S|$ stochastic matrix, denoted M_L, where the (x, y)-th entry in this matrix M_L denotes the probability $P_{L(x)}(y)$.

The result of the t-fold vector-matrix multiplication $D \cdot (M_L)^t$ is a vector of length $|S|$. The component indexed by $x \in S$ indicates the probability that state x is reached after the initial random choice followed by t local search steps. Therefore, the "success probability", the probability of reaching a solution within one run of the REPEAT-loop, is given by

$$p = \sum_{x \in S_0} [D \cdot (M_L)^t]_x$$

If we could calculate or estimate p (which is quite difficult), then an appropriate number of restarts (number of REPEAT-loops) would be c/p for some constant c. This is so, since the probability of missing a solution in each of the c/p trials is

$$(1 - p)^{c/p} \leq (e^{-p})^{c/p} = e^{-c}$$

It is difficult to calculate p since the respective vector D and the matrix M_L are exponential in the size of π, and both, D and M_L, depend individually on the input instance π. We could greatly simplify the analysis, if

- it is possible that the complexity analysis depends merely on the input size $n = |\pi|$, and not on π itself,
- if the number of states of the Markov chain (that is, the vector D, and the matrix M_L) can be reduced.

These requirements can indeed be fulfilled if we divide the search space S (the set of states) into a few categories (equivalence classes), like

$$S = S_0 + S_1 + \cdots + S_n$$

and we base the Markov chain (and its analysis) on these "super states". A quite natural division of S into categories is in terms of the solution distance. Define

$$S_i = \{x \in S \mid d(x) = i\}$$

Example. Continuing the above SAT example it can be seen, that a Markov chain approach using the "super states" $0, 1, \ldots, n$, indicating the solution distance $d(x)$ of the actual assignment x, can be used. Consider the case of k-SAT, and let Y_t be a random variable describing the state number at time t. Then Y_0 is determined by the initial distribution D:

$$P(Y_0 = j) = \binom{n}{j} 2^{-n}$$

The transition probabilities from time t to time $t + 1$ are as follows.

$$P(Y_{t+1} = j - 1 \mid Y_t = j) = \frac{1}{k} \quad \text{and} \quad P(Y_{t+1} = j + 1 \mid Y_t = j) = \frac{k - 1}{k}$$

Let $q(t,j) = P(Y_t = 0 \mid Y_0 = j)$, i.e. the probability that the state 0 (which indicates that a solution is found) is reached after exactly t steps, under the condition that the initial solution distance is j. Clearly, we have $q(t,j) = 0$ for $t < j$. For $t = j$ we have $q(t,j) = (\frac{1}{k})^j$, and in the general case, $t > j$, letting $t = (1+\alpha)j$, we get (using the ballot theorem, cf. [6]),

$$q((1+\alpha)j, j) = \binom{(1+\alpha)j}{\alpha j/2} \cdot \frac{1}{1+\alpha} \cdot \left((\frac{k-1}{k})^{\alpha/2} \cdot (\frac{1}{k})^{1+\alpha/2} \right)^j$$

Asymptotically, and concentrating on the exponential terms, we can approximate this probability as follows.

$$q((1+\alpha)j, j) \approx \left[2^{h(\frac{\alpha}{2(1+\alpha)})(1+\alpha)} \cdot (\frac{k-1}{k})^{\alpha/2} \cdot (\frac{1}{k})^{1+\alpha/2} \right]^j =: \beta^j$$

where h is the binary entropy function [2]. The following figure shows β as a function of α.

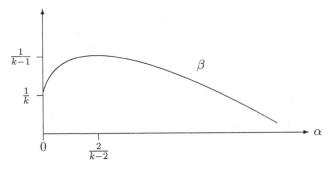

In the beginning (i.e. $\alpha = 0$, that is, $t = j$) we have $\beta = 1/k$, but the highest value, $\beta = 1/(k-1)$ is reached later when $\alpha = 2/(k-2)$, that is, $t = j \cdot k/(k-2)$. We find this quite surprising. It shows an advantage of stochastic local search as compared to deterministic local search, which, in this case would explore the states, starting from the initial state x up to states within Hamming distance j. Then a solution would be found. Doing this systematic search in a backtracking fashion (as in [14,4,5]) requires k^j search steps (= recursive calls of the search procedure), whereas the probability of finding the solution in a single run of stochastic search of at least $j \cdot k/(k-2)$ (not just j) steps is $(1/(k-1))^j$. Using $O((k-1)^j)$ restarts, as explained above, this results in an algorithm which finds the solution (of distance j) almost surely.

4   Restarts

To make the discussion complete, we need to combine this analysis of the local search process (given that the solution distance is j) with the initial random distribution D. Altogether we combine a random initial choice of a solution candidate, followed by at least $j \cdot k/(k-2)$ local search steps according to $L(x)$. Since we don't know the particular j in advance, we can use the upper bound $n \cdot k/(k-2)$ for the number of local search steps before to restart. The resulting success probability is

$$\sum_{j=0}^{n} \binom{n}{j} \cdot 2^{-n} \cdot (\frac{1}{k-1})^{j} = (\frac{k}{2k-2})^{n}$$

by the binomial theorem. Let $r(t)$ be the probability that we reach a solution after exactly t local search steps, following the initial guess of a candidate. Then this function reaches its maximum at $t = n/(k-2)$, where this maximum value conicides, up to polynomial factors, with the above, namely $\beta^{n} = (\frac{k}{2k-2})^{n}$. The following picture indicates the situation.

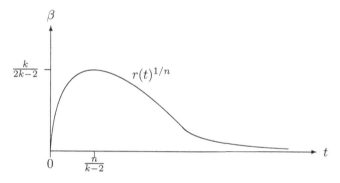

Now it should be clear that it is not the best strategy to make just a single run of the local search algorithm until eventually a solution is reached. Apart from these theoretical results, experiments with this "pure random walksat" procedure, as reported in [12], page 99ff, or in [7], page 258ff, suggest that the algorithm becomes trapped in a "metastable state" for a very long time, and does not reach the solution during this phase. Therefore, a much better strategy is to restart after $t_0 > \frac{n}{k-2}$ steps if no solution was found so far. This strategy results in a picture like the following.

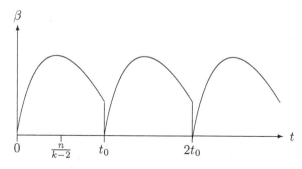

In the first phase, for $0 \leq t \leq t_0$, let the probability of finding a solution be p (up to a polynomial factor we have $p = (\frac{k}{2k-2})^n$). Then the probability of finding a solution, not in the first, but in the second phase, is $(1-p) \cdot p$, and in general, for the i-th phase, $(1-p)^{i-1} \cdot p$. These probabilities are much greater than in the above situation (without any restart), which are, up to some polynomial factor, $r(t_0), r(2t_0), \ldots$.

Usually, when using a Markov chain approach to analyze a random algorithm it is the issue of rapid mixing which is of concern, i.e. reaching the stationary distribution quickly. It is interesting to note that the situation here is quite different. The probability of finding a solution is highest within a certain initial number of steps, afterwards when reaching the stationary distribution, it decreases drastically. Therefore, restarting is an important concept here which, in our opinion, has been neglected so far.

Acknowledgement

I would like to thank Hans Kestler for many productive and intuitive discussions on the subject.

References

1. Aarts, E., Lenstra, J.K.: Local Search in Combinatorial Optimization. Princeton University Press, Princeton, NJ (2003)
2. Ash, R.B.: Information Theory. Dover, Mineola, NY (1965)
3. Bolc, L., Cytowski, J.: Search Methods for Artificial Intelligence. Academic Press, San Diego (1992)
4. Dantsin, E., Goerdt, A., Hirsch, E.A., Schöning, U.: Deterministic algorithms for k-SAT based on covering codes and local search. In: Welzl, E., Montanari, U., Rolim, J.D.P. (eds.) ICALP 2000. LNCS, vol. 1853, pp. 236–247. Springer, Heidelberg (2000)
5. Dantsin, E., Goerdt, A., Hirsch, E.A., Kannan, R., Kleinberg, J., Papadimitriou, C., Raghavan, P., Schöning, U.: A deterministic $(2 - \frac{2}{k+1})^n$ algorithm for k-SAT based on local search. Theoretical Computer Science 289(1), 69–83 (2002)
6. Feller, W.: An Introduction to Probability Theory and Its Applications. Wiley, Chichester (1968)
7. Hartmann, A.K., Weigt, M.: Phase Transitions in Combinatorial Optimization Problems. Wiley-VHC, Chichester (2005)
8. Hoos, H.: Stochastic Local Search - Methods, Models, Applications. PhD dissertation, Universität Darmstadt (1998)
9. Hoos, H., Stützle, T.: Stochastic Local Search - Foundations and Applications. Morgan Kaufmann, San Francisco (2004)
10. Michiels, W., Aarts, E., Korst, J.: Theoretical Aspects of Local Search. Springer, Heidelberg (2007)
11. Papadimitriou, C.H.: On selecting a satisfying truth assignment. In: Proceedings of the 32nd Ann. IEEE Symp. on Foundations of Computer Science, pp. 163–169. IEEE Computer Society Press, Los Alamitos (1991)

12. Perkus, A., Istrate, G., Moore, C. (eds.): Computational Complexity and Statistical Physics. Oxford University Press, Oxford (2006)
13. Schneider, J.J., Kirkpatrick, S.: Stochastic Optimization. Springer, Heidelberg (2006)
14. Schöning, U.: On The Complexity Of Constraint Satisfaction Problems. Ulmer Informatik-Berichte, Nr. 99-03. Universität Ulm, Fakultät für Informatik (1999)
15. Schöning, U.: A probabilistic algorithm for k-SAT and constraint satisfaction problems. In: Proceedings 40th IEEE Symposium on Foundations of Computer Science, pp. 410–414 (1999)
16. Schöning, U.: A probabilistic algorithm for k-SAT based on limited local search and restart. Algorithmica 32(4), 615–623 (2002)

Spatial and Temporal Resource Allocation for Adaptive Parallel Genetic Algorithm

K.Y. Szeto

Department of Physics
Hong Kong University of Science and Technology,
Clear Water Bay, Hong Kong, SAR, China
{phszeto@ust.hk}

Abstract. Adaptive parameter control in evolutionary computation is achieved by a method of computational resource allocation, both spatially and temporally. Spatially it is achieved for the parallel genetic algorithm by the partitioning of the search space into many subspaces. Search for solution is performed in each subspace by a genetic algorithm which domain of chromosomes is restricted inside that particular subspace. This spatial allocation of computational resource takes the advantage of exhaustive search which avoids duplicate effort, and combine it with the parallel nature of the search for solution in disjoint subspaces by genetic algorithm. In each subspace, temporal resource allocation is also made for different competing evolutionary algorithms, so that more CPU time is allocated to those algorithms which showed better performance in the past. This general idea is implemented in a new adaptive genetic algorithm using the formalism of mutation matrix, where the need for setting a survival probability is removed. The temporal resource allocation is introduced and implemented in a single computer using the quasi-parallel time sharing algorithm for the solution of the zero/one knapsack problem. The mutation matrix $M(t)$ is constructed using the locus statistics and the fitness distribution in a population $A(t)$ with N rows and L columns, where N is the size of the population and L is the length of the encoded chromosomes. The mutation matrix is parameter free and adaptive as it is time dependent and captures the accumulated information in the past generation. Two competing strategies of evolution, mutation by row (chromosome), and mutation by column (locus) are used for the competition of CPU time. Time sharing experiment on these two strategies is performed on a single computer in solving the knapsack problem. Based on the investment frontier of time allocation, the optimal configuration in solving the knapsack problem is found.

1 Introduction

Successful applications of genetic algorithm using the Darwinian principle of survival of the fittest have been implemented in many areas [1,2], such as in solving the crypto-arithmetic problem [3], time series forecasting [4], traveling salesman problem [5], function optimization [6], and adaptive agents in stock markets [7,8]. A drawback for the practitioners of genetic algorithm is the need for expertise in the specific

S.G. Akl et al.(Eds.): UC 2007, LNCS 4618, pp. 188–198, 2007.
© Springer-Verlag Berlin Heidelberg 2007

application, as its efficiency depends very much on the parameters chosen in the evolutionary process. An example of this drawback can be found in the ad-hoc manner in choosing the selection mechanism, where different percentages of the population for survival for different problems are generally required. Since the parameters used in a specific application will usually not be optimal in a different application, the traditional usage of genetic algorithm is more like an artisan approach. To tackle this shortcoming, we attempt an adaptive approach where the number of parameters requiring input from experts is reduced to a minimum. The basic aim is to make use of past data collected in the evolutionary process of genetic algorithm to tune the parameters such as the survival probability of a chromosome. In this paper we provide a general framework for adaptive parameter control for parallel genetic algorithm, both spatially and temporally.

Temporally we can collect the statistics of the chromosomes in past generations to decide the probability of mutation of each locus [9]. A general formalism using a mutation matrix has been introduced with generalization to adaptive crossover mechanism reported in a recent publication [10,11], which was reviewed briefly in Section 2. Here we emphasize not only the adaptive parameter control in a particular algorithm of evolutionary computation, but also the competition among several adaptive algorithms for computational time, assuming that there is a single CPU available. This competition of CPU time then addresses the problem of adaptive temporal allocation of resource to individual algorithm by a quasi-parallel process, which is stochastic in nature [12]. To illustrate this problem in layman's terms, let's consider two students, Alice and Bob, working independently on a difficult problem in cryptography. The Professor only provides one computer for them. Without knowing who is the better student, the professor gives Alice and Bob equal time sharing for the use of the computer. However, after some time, Alice shows more promising results. Thus, it is wise for the professor to give her more time to use the computer. A more sophisticated professor will then measure the risk of giving all time to Alice, and adjust the probability of computer usage at a given time by Alice accordingly. The problem of temporal allocation of computer time to Alice and Bob therefore is a problem of portfolio management, when the probability distribution functions of finding a solution by Alice and by Bob should be used in an optimal way. An excellent method in portfolio management has been provided by Markowitz [13] to locate the optimal parameters in running the bottleneck program in a single computer that satisfies the criteria of both high speed and high confidence. These ideas in laymen terms have been implemented technically by Hogg and Huberman and collaborators [14] and Szeto and Jiang [12]. The formalism is basically one of quasi-parallel genetic algorithm, which is aimed at combining existing algorithms into new ones that are unequivocally preferable to any of the component algorithms using the notion of risk in economics [13]. Here we assume that only one computer is available and the sharing of resource is realized only in the time domain. The concept of optimal usage is defined economically by the "investment frontier", characterized by low risk and high speed to solution. We will illustrate this idea in Section 3 where we combine the work on mutation only genetic algorithm and time sharing in the framework of quasi-parallel genetic algorithm. We test this approach on the 0/1 knapsack problem with satisfactory results, locating the investment frontier for the knapsack problem.

Spatially, we consider a simple partitioning of the solution space into disjoint subspaces, so that we can perform parallel search in each subspace with adaptive genetic algorithm with temporal parameter control. Our idea is to combine the advantage of exhaustive search, which does not duplicate effort in search, with parallel genetic algorithm. There will be communication between populations in different subspaces, but there is no exchange of chromosomes and all evolutionary processes are restricted to a given subspace. In this way, the dimensionality of the solution space is reduced. However, this is the trivial advantage of spatial partition. The real advantage comes from the allocation of computational resource to those populations which in the past show more promise in finding the solution. In layman's term, let's consider the search for a lost key in the beach. One first partitions the beach area into giant grids, and perform parallel intelligent search in a given grid with the help of friends. By looking for characteristic traces of the lost key, one naturally should spend more time in searching in a grid where there are more traces. In Section 4, we put this simple idea in more technical terms and suggest a way for performance evaluation for the search process in individual subspace.

2 Mutation Matrix

In traditional simple genetic algorithm, the mutation/crossover operators are processed on the chromosome indiscriminately over the loci without making use of the loci statistics, which has been demonstrated to provide useful information on mutation operator [9]. In our mutation matrix formalism, the traditional genetic algorithm can be treated as a special case. Let's consider a population of N chromosomes, each of length L and binary encoded. We describe the population by a NxL matrix, with entry $A_{ij}(t), i = 1,..., N; j = 1,..., L$ denoting the value of the jth locus of the ith chromosome. The convention is to order the rows of A by the fitness of the chromosomes, $f_i(t) \le f_k(t) \; for \; i \ge k$. Traditionally we divide the population of N chromosomes into three groups: (1) Survivors who are the fit ones. They form the first N_1 rows of the population matrix $A(t+1)$. Here $N_1 = c_1 N$ with the survival selection ratio $0 < c_1 < 1$. (2) The number of children is $N_2 = c_2 N$ and is generated from the fit chromosomes by genetic operators such as mutation. Here $0 < c_2 < 1 - c_1$ is the second parameter of the model. We replace the next N_2 rows in the population matrix $A(t+1)$. (3) the remaining $N_3 = N - N_1 - N_2$ rows are the randomly generated chromosomes to ensure the diversity of the population so that the genetic algorithm continuously explores the solution space.

In our formalism, we introduce a mutation matrix with elements $M_{ij}(t) \equiv a_i(t)b_j(t), i = 1,..., N; j = 1,..., L$ where $0 \le a_i(t), b_j(t) \le 1$ are called the row mutation probability and column mutation probability respectively. Traditional genetic algorithm with mutation as the only genetic operator corresponds to a time *independent* mutation matrix with elements $M_{ij}(t) \equiv 0$ for $i = 1,..., N_1$, $M_{ij}(t) \equiv m \in (0,1)$ for

$i = N_1 + 1, ..., N_2$, and finally we have $M_{ij}(t) \equiv 1$ for $i = N_2 + 1, ..., N$. Here m is the time independent mutation rate. We see that traditional genetic algorithm with mutation as the only genetic operator requires at least three parameters: N_1, N_2, and m.

We first consider the case of mutation on a fit chromosome. We expect to mutate only a few loci so that it keeps most of the information unchanged. This corresponds to ``*exploitation*'' of the features of fit chromosomes. On the other hand, when an unfit chromosome undergoes mutation, it should change many of its loci so that it can explore more regions of the solution space. This corresponds ``*exploration*''. Therefore, we require that $M_{ij}(t)$ should be a monotonic increasing function of the row index i since we order the population in descending order of fitness. There are many ways to introduce the row mutation probability. One simple solution is to use $a_i(t) = (i-1)/(N-1)$. Next, we must decide on the choice of loci for mutation once we have selected a chromosome to undergo mutation. This is accomplished by computing the locus mutation probability of changing to X (X=0 or 1) at locus j as p_{jX} by

$$p_{jX} = \sum_{k=1}^{N} (N + 1 - k) \delta_{kj}(X) \bigg/ \sum_{m=1}^{N} m \tag{1}$$

Here k is the rank of the chromosome in the population. $\delta_{kj}(X) = 1$ if the *j-th* locus of the *k-th* chromosome assume the value X, and zero otherwise. The factor in the denominator is for normalization. Note that p_{jX} contains information of both locus and row and the locus statistics is biased so that heavier weight for chromosomes with high fitness is assumed. This is in general better than the original method of Ma and Szeto[9] where there is no bias on the row. After defining p_{jX}, we define the column mutation rate as

$$b_j = \left(1 - | p_{j0} - 0.5 | - | p_{j1} - 0.5 |\right) \bigg/ \sum_{j'=1}^{L} b_{j'} \tag{2}$$

For example, if 0 and 1 are randomly distributed, then $p_{j0} = p_{j1} = 0.5$. We have no useful information about the locus, so we should mutate this locus, and $b_j = 1$. When there is definitive information, such as when $p_{j0} = 1 - p_{j1} = 0$ or 1, we should not mutate this column and $b_j = 0$.

Once the mutation matrix **M** is obtained, we are ready to discuss the strategy of using **M** to evolve **A**. There are two ways to do Mutation Only Genetic Algorithm. We can first decide which row (chromosome) to mutate, then which column (locus) to mutate, we call this particular method the *Mutation Only Genetic Algorithm by Row* or abbreviated as MOGAR. Alternatively, we can first select the column and then the row to mutate, and we call this the *Mutation Only Genetic Algorithm by Column* or abbreviated as MOGAC.

For MOGAR, we go through the population matrix A(t) by row first. The first step is to order the set of locus mutation probability $b_j(t)$ in descending order. This ordered set will be used for the determining of the set of column position (locus) in the mutation process. Now, for a given row i, we generate a random number x. If $x < a_i(t)$, then we perform mutation on this row, otherwise we proceed to the next row and $A_{ij}(t+1) = A_{ij}(t)$, $j = 1,...,L$. If row i is to be mutated, we determine the set $R_i(t)$ of loci in row i to be changed by choosing the loci with $b_j(t)$ in descending order, till we obtain $K_i(t) = a_i(t)*L$ members. Once the set $R_i(t)$ has been constructed, mutation will be performed on these columns of the i-th row of the A(t) matrix to obtain the matrix elements $A_{ij}(t+1)$, $j = 1,...,L$. We then go through all N rows, so that in one generation, we need to sort a list of L probabilities $b_j(t)$ and generate N random numbers for the rows. After we obtained A(t+1), we need to compute the $M_{ij}(t+1) = a_i b_j(t+1)$ and proceed to the next generation.

For MOGAC, the operation is similar to MOGAR mathematically except now we rotate the matrix A by 90 degrees. Now, for a given column j we generate a random number y. If $y < b_j(t)$, then we mutate this column, otherwise we proceed to the next column and $A_{ij}(t+1) = A_{ij}(t)$, $i = 1,...,N$. If column j is to be mutated, we determine the set $S_j(t)$ of chromosomes in column j to be changed by choosing the rows with the $a_i(t)$ in descending order, till we obtain $W_j(t) = b_j(t)*N$ members. Since our matrix A is assumed to be row ordered by fitness, we simply need to choose the $N, N-1,..., N-W_j+1$ rows to have the j-th column in these row mutated to obtain the matrix elements $A_{ij}(t+1)$, $i = 1,...,N$. We then go through all L columns, so that in one generation, we need to sort a list of N fitness values and generate L random numbers for the columns.

3 Temporal Allocation of Resource

Our mutation matrix is adaptive and time dependent. The two strategies of using the mutation matrix formalism, MOGAR and MOGAC, present a good example for temporal allocation of computational resource. The framework for proper mixing of computing algorithms is the quasi-parallel algorithm of Szeto and Jiang [12]. Here we first summarize this algorithm. A simple version of our quasi-parallel genetic algorithm ($QPGA = (M, SubGA, \Gamma, T)$) consists of M independent sub-algorithms $SubGA$. The time sharing of the computing resource is described by the resource allocation vector Γ. If the total computing resource is R, shared by M sub-algorithms G_i, $i=1,2,...,M$, with resource R_i assigned to G_i in unit time, then we introduce

$\tau_i = R_i / R$, $0 \le \tau_i \le 1$ for any $i=1,2,...,M$, and $\sum_{i=1}^{M} \tau_i = 1$, and the allocation of resource for sub-algorithms is defined by the *resource allocation vector*, $\Gamma = (\tau_1, \tau_2, \cdots \tau_M)'$. In our case, we have $M=2$ and our *SubGA* are MOGAR and MOGAC. For resource allocation, we only have one parameter $0 < \gamma < 1$ which is the fraction of time the computer is using MOGAR, and the remaining fraction of time $(1-\gamma)$ we use MOGAC. The termination criterion T is used to determine whether a QPGA should stop running. Thus, we have a mixing of MOGAR with MOGAC in the framework of quasi-parallel genetic algorithm with a mixing parameter γ. The parallel genetic algorithm described above can be implemented in a serial computer. For a particular generation t, we will generate a random number z. If $z < \gamma$, then we perform MOGAR, otherwise we perform MOGAC to generate the population $A(t+1)$. We now apply this quasi-parallel mutation only genetic algorithm to solve the 0/1 knapsack problem and try to obtain the investment frontier that give the mixing parameter γ that yields the fastest speed in solving the problem while also running with most certainty (minimum risk) of getting the solution.

4 Spatial Allocation of Resource

Let's begin with the simple case of binary encoding of the solution of a typical search problem for the maximum of a function. The quality of the solution is defined by the length L of the chromosomes. In exhaustive search, we simply try all 2^L chromosomes in the solution space. If we partition the solution space into n subspaces, $\left\{ n = 2^k, (0 < k < L) \right\}$, we can perform intelligent search in each of these n subspaces, so that there is an effective reduction in the dimension of search space from 2^L to 2^{L-k}. For example, we use n=4 and use the first two loci of the chromosomes: $S_{cd} \equiv \left\{ (cdx_{k+1}...x_L), x_i \in (0,1); i = k+1,...,L; (c,d \in (0,1)) \right\}$. The four sets of chromosomes $(S_{00}, S_{01}, S_{10}, S_{11})$ describe typical solution restricted to the four subspaces labeled by the first two loci. These subspaces are mutually exclusive so that search in each subspace can be performed independently without duplication of effort that often appears had the search not been restricted to individual subspace. Operationally, one can search sequentially or randomly in these subspaces. For illustrative purpose, we assume sequentially search in the order of $(S_{00}, S_{01}, S_{10}, S_{11})$ by assigning four populations $(A_{00}, A_{01}, A_{10}, A_{11})$ correspondingly for each of the subspace. In order to implement adaptive spatial allocation of resource, we evolve each population (A_{cd}) for some generations (G_{cd}). A time sharing process for the evolution of many populations with one CPU can then be implemented on these four populations for the four subspaces. As in Section 3, we can do the allocation of CPU by the following steps:

1. Divide searching into segments with equal length τ which denotes the total number of generations the program will evolve in all subspaces in the given segment.

2. Initialize each population with $\left(G_{00}(0) = G_{01}(0) = G_{10}(0) = G_{11}(0) = \tau/n \right)$.

3. The population which represents a subspace with a higher probability to contain the global optimal solution will gain more computational time. The adaptive change of the sharing of resource can be done with a measure of the population fitness $\left(F_{00}, F_{01}, F_{10}, F_{11} \right)$ of $\left(A_{00}, A_{01}, A_{10}, A_{11} \right)$.

In this example, we need to define the population fitness, as well as a partition of the segment into n pieces for the n population to evolve in the next segment. Given a measure of population fitness, we can partition the segment by the monotonic increasing function $\left(P_{cd}(F_{cd}) \right)$ of population fitness so that the division of the segment time is given by

$$G_{cd}(t+1) = \tau \frac{P_{cd}(F_{cd}(t))}{P_{00}(F_{00}(t)) + P_{01}(F_{01}(t)) + P_{10}(F_{10}(t)) + P_{11}(F_{11}(t))} \tag{3}$$

Here t is the label for the segment and initially, $P_{cd} = 1/n$.

There are two remaining problems to be solved before one can use spatial resource allocation. The first one concerns the function $P_{cd}(F_{cd})$. If we consider the evolving populations $\left(A_{00}, A_{01}, A_{10}, A_{11} \right)$ representing states of a large system in different configuration in the space of solutions, then they may be considered as the state of the system with specific "energy", represented by their fitness values $\left(F_{00}, F_{01}, F_{10}, F_{11} \right)$. The similarity of an evolving physical system with changing energy landscape leads naturally to the Boltzmann form of $P \propto e^F$. The partitioning of the segment τ accordingly follows the partition of unity of the Boltzmann factor in the canonical ensemble in statistical mechanics. Although this choice of the P is a heuristic one, it has the advantage of the interpretation of probability. However, even without appealing to physics, this choice for P in practice has the advantage of amplifying the effect of difference in population fitness, which is usually not large for practical size of segment τ.

The second problem concerns the choice of F. Here we define F as the fitness of the population A in a given subspace S. For each member of the population A, which is a chromosome c, one can associate with c a fitness f. This fitness f is the usual definition of fitness in single population genetic algorithm. For example, if the problem is to find the maximum of a given function H(x), then a trial solution $\tilde{x}$ can be represented as a binary string, which corresponds to the chromosome c in genetic algorithm. The fitness f of the chromosome can simply be $f = H(\tilde{x})$. The population fitness for a collection of N chromosomes can then be the average fitness of these N trial solutions. However, a better choice for population fitness is the maximum fitness of the N solutions: $F = \max\{f_i = H(x_i); i = 1,...,N\}$. This is a better choice for population fitness because our objective is to find the maximum of the H(x). In terms of performance evaluation, we can pose the following question: are we spending the effort to find the solution in the right place? By right place, we mean the subspace that actually contains the global maximum of H(x). Supposing that we partition the solution space into four subspaces $\left(S_{00}, S_{01}, S_{10}, S_{11}\right)$ and that the global maximum of H(x) is in S_{cd}. If $G_{cd} > \langle G \rangle$, where $\langle G \rangle = \frac{1}{n}\left(\sum_{c'd'} G_{c'd'}\right)$ is the average genera-

tion, then we can say that we are correctly devoting our search effort in the right place. Over a large number of numerical experiments that run this search for the global maximum of H(x), the probability Q that $G_{cd} > \langle G \rangle$ can be computed. This probability Q represents the probability of correct spatial resource allocation in the search for maximum H(x) and it can be used as a measure of performance of the algorithm for the given choice of P and F.

5   The Zero/One Knapsack Problem

The model problem to test our ideas on resource allocation for adaptive genetic algorithm is the Knapsack problem. We define the 0/1 knapsack problem [15] as follow. Given L items, each with profit P_i, weight w_i and the total capacity limit c, we need to select a subset of L items to maximize the total profit, but its total weight does not exceed the capacity limit.   Mathematically, we want to find the set $x_i \in \{0,1\}, i = 1,...,L$ to

$$\text{Maximize} \sum_{j=1}^{L} P_j x_j \text{ subjected to constraint } c \geq \sum_{j=1}^{L} w_j x_j \qquad (4)$$

We consider a particular knapsack problem with size $L=150$ items, c=4000, $P_j \in [0,1000]$, and $w_j \in [0,100]$ chosen randomly but afterwards fixed. We have found evidence that mixing MOGAR with MOGAC in a time-sharing manner produces superior results compared to dynamic programming in numerical experiment for the 0/1 knapsack problem. More importantly is the finding of the optimal time sharing parameter by locating the investment frontier of our mutation only genetic algorithm.

Our choice of the mutation matrix is the simple form of $a_i(t)=(i-1)/(N-1)$. Here N (=100) is the size of the population in all genetic algorithms. For a given generation time t, we generate a random number z. If $z < \gamma$, then we perform MOGAR, otherwise we perform MOGAC to generate the population $A(t+1)$. Next we address the stopping criterion. We use exhaustive search to locate the true optimal solution of the knapsack problem. Then, we run our mixed MOGA in temporal resource allocation as

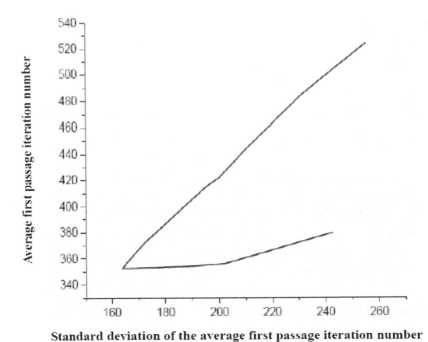

Standard deviation of the average first passage iteration number

Fig. 1. Illustration of the investment frontier in temporal allocation of resource for using MOGAR and MOGAC as competing algorithms in the search for solution of the 0/1 Knapsack problem. Different point on the curve corresponds to different time sharing parameter γ. The optimal allocation is defined by the tip of the curve, which corresponds to the time sharing factor at $\gamma_c = 0.22 \pm 0.02$.

defined in section 3 for 1000 times to collect statistical data. For each run, we define the stopping time to be the generation number when we obtain the optimal solution, or 1500, whichever is smaller. The choice of 1500 as the upper bound is based on our numerical experience since for a wide range of γ, all the runs are able to find the optimal solution within 1500 generations. Only for those extreme cases where γ is near 1 or 0, meaning that we use MOGAR alone or MOGAC alone, a few runs fail to find the optimal solution within 1500 generations. This is expected since we know row mutation only GA has low speed of convergence while column mutation only GA has early convergence problem. These extreme cases turn out to be irrelevant in our search of the investment frontier. When we plot the mean first passage time to solution and its standard deviation of 1000 runs as a function of the time sharing parameter γ, we found the investment frontier shown in Fig.1. The curve is parameterized by γ. We see that there is a point on the curve which is closest to the origin. This point is unique in this experiment, corresponding to a value of $\gamma_c = 0.22 \pm 0.02$. The interpretation of this time sharing parameter is that our temporal allocation produces the fastest and most reliable (least risky) hybrid algorithm in finding the optimal solution of the 0/1 knapsack problem. In another word, the investment frontier of this problem consists of a single critical point at γ_c.

For spatial resource allocation, we also use the knapsack problem as benchmark test. We evaluate our algorithm for L=15,20,25, and we use τ =80 and divide our solution space into four subspaces. The size of each of the four populations is 25 and simple MOGAR is used for the evolution in each subspace. The probability Q of correct spatial resource allocation in the search for solution of the knapsack problem of size L are (0.687,0.949,0.972) for L=15,20,25 respectively.

6 Discussion

The simple ideas illustrated with a layman's example in the introduction is formalized and tested with the 0/1 knapsack problem in this paper. The general idea of spatial and temporal resource allocation in adaptive parallel genetic algorithm can be extended to include crossover [11] and more complex choices of competing algorithms. One of the possible extensions for spatial resource allocation is to do an intelligent partition of the solution space initially, so as to speed up the search. In this paper we only use the first two loci for the partitioning of the solution space into four subspaces. In general, we should ask what are the best choices for these two loci and when should we perform the partitioning. Should we do it repetitively, so that the search becomes some kind of iterative partitioning with a fast shooting at the location of global optimum? For temporal allocation, we can attempt a more sophisticated analysis of the investment frontier when there are more than two competing algorithms. In our present example, the frontier consists of a single critical point, representing the critical value of time sharing of mixing MOGAR and MOGAC to be 0.22, meaning that statistically we should use 22% of the computational resource on mutation by row, and 78% on mutation by column when solving the knapsack problem. In general, the frontier is a critical surface which topology should be of interest for further theoretical analysis.

Acknowledgement

K.Y. Szeto acknowledged that this work is supported by RGC grant no. HKUST 603203 and DAG04.05/SC23 Numerical help from Zhao S.Y. is acknowledged.

References

1. Holland, J.H.: Adaptation in Natural and Artificial Systems. University of Michigan Press, Ann Arbor, MI (1975)
2. Goldberg, D.E.: Genetic Algorithms in Search, Optimization, and Machine Learning. Addison-Wesley, Reading, MA (1989)
3. Li, S.P., Szeto, K.Y.: Crytoarithmetic problem using parallel Genetic Algorithms, Mendl'99, Brno, Czech. (1999)

4. Szeto, K.Y., Cheung, K.H.: Multiple time series prediction using genetic algorithms optimizer. In: IDEAL'98. Proceedings of the International Symposium on Intelligent Data Engineering and Learning, Hong Kong, pp. 127–133 (1998)
5. Jiang, R., Szeto, K.Y., Luo, Y.P., Hu, D.C.: Distributed parallel genetic algorithm with path splitting scheme for the large traveling salesman problems. In: Shi, Z., Faltings, B., Musen, M. (eds.) Proceedings of Conference on Intelligent Information Processing, 16th World Computer Congress 2000, Beijing, August 21-25, 2000, pp. 478–485. Publishing House of Electronic Industry (2000)
6. Szeto, K.Y., Cheung, K.H., Li, S.P.: Effects of dimensionality on parallel genetic algorithms. In: Proceedings of the 4th International Conference on Information System, Analysis and Synthesis, Orlando, Florida, USA, vol. 2, pp. 322–325 (1998)
7. Szeto, K.Y., Fong, L.Y.: How adaptive agents in stock market perform in the presence of random news: a genetic algorithm approach. In: Leung, K.-S., Chan, L., Meng, H. (eds.) IDEAL 2000. LNCS, vol. 1983, pp. 505–510. Springer, Heidelberg (2000)
8. Fong, A.L.Y., Szeto, K.Y.: Rule Extraction in Short Memory Time Series using Genetic Algorithms. European Physical Journal B 20, 569–572 (2001)
9. Ma, C.W., Szeto, K.Y.: Locus Oriented Adaptive Genetic Algorithm: Application to the Zero/One Knapsack Problem. In: Proceeding of The 5th International Conference on Recent Advances in Soft Computing, RASC2004, Nottingham, UK, pp. 410–415 (2004)
10. Szeto, K.Y., Zhang, J.: In: Lirkov, I., Margenov, S., Waśniewski, J. (eds.): LSSC 2005. LNCS, vol. 3743, pp. 189–196. Springer, Heidelberg (2006)
11. Law, N.L., Szeto, K.Y.: Adaptive Genetic Algorithm with Mutation and Crossover Matrices. In: Proceeding of the 12th International Joint Conference on Artificial Intelligence (IJCAI-07) January 6 - 12, 2007 (Volume II) Theme: AI and Its Benefits to Society, Published by International Joint Conferences on Artificial Intelligence, IJCAI-07. Hyderabad, India, pp. 2330–2333 (2007)
12. Szeto, K.Y., Rui, J.: A quasi-parallel realization of the Investment Frontier in Computer Resource Allocation Using Simple Genetic Algorithm on a Single Computer. In: Fagerholm, J., Haataja, J., Järvinen, J., Lyly, M., Råback, P., Savolainen, V. (eds.) PARA 2002. LNCS, vol. 2367, pp. 116–126. Springer, Heidelberg (2002)
13. Markowitz, H.: J. of Finance 7, 77 (1952)
14. Huberman, B.A., Lukose, R.M., Hogg, T.: An economics approach to hard computational problems. Science 275(3), 51–54 (1997)
15. Gordon, V., Bohm, A., Whitley, D.: A Note on the Performance of Genetic Algorithms on Zero-One Knapsack Problems. In: Proceedings of the 9th Symposium on Applied Computing (SAC'94), Genetic Algorithms and Combinatorial Optimization, Phoenix, Az, pp. 194–195 (1994)

Gravitational Topological Quantum Computation

Mario Vélez and Juan Ospina

Logic and Computation Group
School of Sciences and Humanities
EAFIT University
Medellin, Colombia
{mvelez,jospina}@eafit.edu.co

Abstract. A new model in topological quantum computing, named Gravitational Topological Quantum Computing (GTQC), is introduced as an alternative respect to the Anyonic Topological Quantum Computing and DNA Computing. In the new model the quantum computer is the quantum space-time itself and the corresponding quantum algorithms refer to the computation of topological invariants for knots, links and tangles. Some applications of GTQC in quantum complexity theory and computability theory are discussed, particularly it is conjectured that the Khovanov polynomial for knots and links is more hard than #P-hard; and that the homeomorphism problem, which is non-computable, maybe can be computed after all via a hyper-computer based on GTQC.

Keywords: Gravitational computer, Topological quantum computing, link invariants, entanglement, complexity, computability.

1 Introduction

There are many practical problems whose solutions can not be determined using a classical computer [1]. Actually is believed that such hard problems can be solved using quantum computers [2]. More over was recently proposed that these hard problems could be solved using topological quantum computers, specifically anyonic topological quantum computers [3]. In this model, the quantum computer is a certain condensed matter system called anyon system and the computations are realized using braiding of anyon lines.

In this work we intend to describe a new model of topological quantum computer, which is an alternative respect to the anyonic model. The new model is named gravitational topological quantum computer and in this case the quantum computer is the quantum space-time itself. It is expected that the GTQC will be able to solve hard problems and will be possible to think about gravitational topological quantum hyper-computers which are able to solve incomputable problems.

Very recently the topological quantum computation has been oriented to the computation of invariant polynomials for knots and links, such as Jones, HOMFLY, Kauffman polynomials [4]. The physical referent that is keeping on mind in this application of TQC is certain kinds of anyon structures which are realized as

S.G. Akl et al.(Eds.): UC 2007, LNCS 4618, pp. 199–213, 2007.
© Springer-Verlag Berlin Heidelberg 2007

condensed matter systems. More over any quantum computation can be reduced to an instance of computation of Jones polynomial for determined knot or link [5,3].

From the other side, it is known that link invariants also appear in quantum gravity as the amplitudes for certain gravitational process . Specifically the Jones, HOMFLY and Kauffman polynomials can be obtained as the vacuum states of the quantum gravitational field [6]. In other words the quantum gravity is able to compute Jones polynomials.

For hence from the perspective of the TQC it is possible to think in other computational model which can be named Gravitational Topological Quantum Computation (GTQC). In GTQC the physical quantum computer is not more an anyon system. In GTQC the quantum computer is the space-time itself; the computer is the quantum gravitational field.

In this work we develop the idea of GTQC and we derive a gravitational quantum algorithm for the approximated calculation of the Jones polynomials, Kauffman bracket and Khovanov homology [7]. Also we discuss the possible applications of GTQC to the quantum complexity theory and to the computability theory.

The more famous topological invariants for knots and links are the following: the Alexander polynomials (1923), the Alexander-Conway polynomials (1960), the Jones polynomials(1984), the HOMFLY polynomials (1986), the bracket polynomials (1987), the Kauffman polynomials (1990), the Vassiliev invariants (1988), the Konsetvich invariants (1990) and the Khovanov cohomology (2000). At general the effective calculations of all these topological invariants are very hard and complex as for to be realized using a classical or traditional computer. Recently it was proposed that these calculations could be efficiently realized using certain kinds of non-classical computers such as the quantum computers and the DNA computers. For hence the aim of the present paper is to show that the gravitational topological quantum computers in a very natural way are able to compute such topological invariants.

2   AJL Algorithm in Anyonic TQC

In [8] Witten was able to obtain the Jones polynomial within the context of the topological quantum field theory of the Chern-Simons kind. Nearly immediately other polynomials were obtained such as the HOMFLY and the Kauffman polynomials [9]. The main idea consists in that the observables (vacuum expectation values) of the TQFT's are by itself topological invariants for knots, links, tangles and three-manifolds.

More in details, given an oriented knot or link and an certain representation of the gauge group, it is possible to build a Wilson Loop operator (colored Wilson Loop) for which the vacuum value is precisely the Jones, HOMFLY and Kauffman polynomials.

Now in [10], Freedman, Kitaev, Larsen and Wang give us a new model of quantum computation, which is named topological quantum computation and given that this model is presented as realized using anyons systems, the new model is renamed as anyonic topological quantum computation [11]. Two important consequences from the model of Freedman *et al* are that many TQFT's can be simulated using a standard quantum computer and that reciprocally a quantum computer can be implemented as an anyonic topological quantum computer. The model of Freedman *et al* is based on directly over TQFT and it involves complicated developments in algebraic topology which are very difficult to understand from the perspective of the computer science.

Fortunately very recently a thanks to the work of Kauffman and Lomonaco [12], a new direction in anyonic topological quantum computation was formulated. This line Kauffman-Lomonaco is based on geometric topology and representation theory and it is more accessible for the computer scientists.

Justly inside of this line of geometric topology and representation theory, and again very recently, a topological quantum algorithm for the approximated computation of the Jones polynomial was discovered by Aharanov, Jones and Landau [4]. This algorithm was its origin in the Freedman-Kitaev-Larsen-Wang model inside of which certain algorithm for the Jones polynomial was in latent state [3,10]. The algorithm AJL was generalized in [5] for the case of HOMFLY polynomial and its structure is discussed in a pedagogical way in [13].

Here we describe the basic ingredients of the AJL algorithm which are: 1) The definition of the Kauffman bracket, 2) the definition of the Jones polynomial in terms of the Kauffman bracket, 3) the representation of the Kauffman bracket as the Markov trace of an unitary representation of the braid algebra via the Temperley-Lieb algebra ; 4) the quantum compilation of the unitary braid representation; and, 5) the Hadamard test.

The Kauffman Bracket for the link L is defined as the following state summation [9,4,5]

$$\langle L \rangle (A, B, J) = \sum_{\alpha} \langle L, \alpha \rangle (A, B) \; J^{-1+k(\alpha)} \tag{1}$$

Where the summation is over every state α which is a representation of every possible smoothing of the link L; A, B and J are the skein variables; $k(\alpha)$ is the number of loops in the state α; and $\langle L, \alpha \rangle (A,B)$ is the contribution of the state α in terms of the skein variables A and B, being the contribution of the loops of the state α given by $J^{(k(\alpha)-1)}$.

The computation of the Kauffman Bracket for a given link L with a high number of crossings and with a very high number of possible states of smoothing, is a hard problem, concretely is a #P-hard problem and for then it can not be solved algorithmically in a polynomial time [4,5].

For the case of very little knots it is possible to use computer algebra software with the aim to compute the corresponding Kauffman brackets. Specifically it is possible to use a program in Mathematica, which was constructed by the mathematician Dror Bar Natan [14]. The input for this program is a representation of the link as a product of the symbols for every crossing, then every crossing is smoothed using the Kauffman skein relation and the product of symbols is expanded and simplified according with the relations among the skein parameters A, B and J.

The Kauffman bracket given by (1) can be normalized to [9,4,5]

$$K(L, A, B, J) = (-A^3)^{-w(L)} \langle L \rangle (A, B, J) \tag{2}$$

Where w(L) is the writhe of L it is to say the sum of the crossing signs of L.

As is well known, the bracket polynomial (1) is invariant under Reidemeister moves II and III but it is not invariant under Reidemeister move I. But the normalized Kauffman bracket (2) is invariant under all the Reidemeister moves and when certain change of variable is made, the Jones polynomial is obtained, namely

$$V(L,t) = \lim_{A \to t^{-\frac{1}{4}}} K(L, A, \frac{1}{A}, -\frac{1}{A^2} - A^2)$$ (3)

where t is the variable of the Jones polynomial. It is worthwhile to note here that the Jones polynomial was discovered by V. Jones [15] using a different method which does not involve the Kauffman bracket.

As we said in the introduction the anyonic topological quantum computation was formulated originally inside the domain of the TQFT's of the algebraic topology. Also we said that Kauffman and Lomonaco were able to construct the ATQC inside the domain of the geometric topology.

The geometric topology is concerned with the study of the geometrical figures which exhibit topological entanglement [16], such as braids, knots, links, tangles and so on. The main idea in geometric topology is to discover algebraic structures which are naturally associated with the topological entanglement. There many of these algebras, some examples are the braid algebra, the Temperley-Algebra, the Iwahori-Hecke algebra, the Birman-Wenlz-Murakami algebra, and their generalizations [17]. To understand the AJL algorithm is necessary to introduce the braid and TL algebras.

The Artin braid [18] on n strands, denoted B_n is generated by elementary braid operators denoted $\{\sigma_1, \sigma_2, , \sigma_{n-1}\}$ with the relations

$$\sigma_i \sigma_j = \sigma_j \sigma_i \text{ for } |i - j| > 1$$ (4)

$$\sigma_i \sigma_{i+1} \sigma_i = \sigma_{i+1} \sigma_i \sigma_{i+1} \text{ for } i = 1, \cdots n - 2.$$ (5)

A generic braid is a word in the operators σ's.

The equation (5) is the Yang-Baxter equation and its solutions are by themselves very important for quantum computing [12]. The group B_n is an extension the symmetric group S_n and as was discovered very recently, B_n has a very rich representation theory including dense sets of unitary transformations [3,10,4,5] and which again is the fundamental importance for the quantum computation in particular and the quantum informatics in general.

Now intimately linked with the braid algebra, appears the algebra named Temperley-Lieb [19], which is denoted $TL_n(J)$ and is generated by the operators $\{I_n, U_1, U_2, \cdots, U_{n-1}\}$ with the relations

$$U_i^2 = J U_i$$ (6)

$$U_i U_{i+1} U_i = U_i$$ (7)

$$U_i U_j = U_j U_i: |i - j| > 1.$$ (8)

The Temperley-Lieb algebra can be represented diagrammatically using certain pictures which are named Kauffman diagrams [20]. It is worthwhile to note here that both the braids and the Kauffman diagrams are particular cases of certain more general structures of topological entanglement which are named tangles. All this means that both the braid algebra B_n and the Temperley-Lieb algebra TL_n are particular cases of a more general algebra which is named tangle algebra. From the other side is very remarkable that the tangle algebra has a very important role in quantum gravity [21] and this fact can be exploited in the sense of the theoretical possibility of a gravitational topological quantum computation, as will be seen more later. Finally it is necessary to have in mind that all knot or link can be obtained as the closure of a given braid or tangle, being a closure the object that is obtained when the extremes of the braid or tangle are identified [4,5].

The topic of the unitary representations of the braid algebra is crucial both for the AJL algorithm in particular as for the anyonic topological quantum computation in general [3,10,4,5,11,12].

There are many ways to obtain such unitary braid representations: 1) Unitary solutions of the Yang-Baxter equations and Yang-Baxterization [12,22]; 2) Unitary representations of the braid algebra on the Temperley-Lieb algebra [3,10,4,5,13]; and 3) The Kauffman's Temperley-Lieb recoupling theory [11]. For the AJL algorithm the more appropriate is the method 2) using the representations of B_n on $TL_n(J)$.

An unitary braid representation can be obtained via TL algebra as: [4,5,11,13]

$$\Phi(\sigma_i) = AI + BU_i \qquad (9)$$

The equation (9) gives the representation for the generators of B_n. For the case of a generic braid b described by certain word in B_n, the equation (9) implies that [11]

$$\Phi(b) = A^{e(b)} I + \Psi(b) \qquad (10)$$

Where $e(b)$ is sum of the exponents in the word that represents b and $\Psi(b)$ is a sum of products in the operators U_i.

Now we introduce certain trace for the representation (9) which is named a Markov trace [4,5,13,11,12]. This trace is similar to usual trace for matrices but the difference is that the Markov trace is defined for the algebras of geometric topology such as Temperley-Lieb, Iwahori-Hecke and Birman-Wenzl-Murakami [17].

The crucial fact here is that the Kauffman bracket for the knot or link that results from the closure of b (denoted $C(b)$), is justly the Markov trace for the representation (10), formally

$$<C(b)>(A,B,J) = tr(\Psi(b)) + A^{e(b)} J^d \qquad (11)$$

Where d is the dimension of the unitary braid representation (9). When $\Psi(b)$ is solved in (10) and the result is substituted in (11) we obtain that

$$<C(b)>(A,B,J) = tr(\Phi(b)) + A^{e(b)} (-d + J^d) \qquad (12)$$

As it is observed in (12), the question to computing the Kauffman bracket for the closure of the braid b is mathematically reduced to the problem of computing the Markov trace of the unitary operator $\Phi(b)$ defined in $TL_n(J)$ [4,5,11,13].

As a particular case, the equation (12) implies that due to (2) and (3), the computation of the Jones polynomial for $C(b)$ is mathematically reduced to the computation of the Markov trace for the unitary braid representation $\Phi(b)$.

The AJL algorithm consists in a method to compute $tr(\Phi(b))$ using a standard quantum computer. The AJL algorithm is realized in two steps, the first step is the compilation process and the second step is execution [13]. The compilation phase consists in translate to the quantum machine language (qbits , two-qbits quantum gates and quantum circuits), the unitary braid representation Φ(b). The execution phase is simply the application of the Hadamard test to the quantum circuit that represents the unitary braid operator Φ(b), being the Hadamard test certain quantum algorithm for determination of the diagonal elements of a unitary matrix. More details for the AJL algorithm are given in [4,5,13,11]. In [5] Wocjan and Yard present a generalization of the AJL algorithm, which is able to compute de HOMFLY polynomial of a given link. The algorithm proposed by Wocjan and Yard is based on directly over the Iwahori-Hecke algebra [5,17] with the corresponding unitary braid representations and the appropriate Markov trace [5,17]. As a particular case the AJL algorithm is obtained given that the IH algebra contains the TL algebra or in other words the HOMFLY polynomial contains as a limit case the Jones polynomial.

Other kinds of algorithms for the computation of Jones polynomials are given in [23].

From the physical point of view, the AJL algorithm may be realized using anyon systems or more exotic systems in condensed matter physics [24].

3 Gravitational Topological Quantum Computation

In this section the idea of a GTQC is introduced. Intuitively a GTCQ is a model of computation using the quantum gravitational field as a quantum computer.

Quantum gravity is an attempt to put together the quantum mechanics and the general relativity according to a single consistent theory [25]. There are actually various models of quantum gravity. From a side we have the string quantum gravity (Schwarz); and from the other side we have the loop quantum gravity (Ahstekar, Smolin, Rovelli, Griego, Gambini, Pullin, Baez, Kauffman) [26,25,6,21] and the spin network quantum gravity (Penrose) From the perspective of the TQC the more adequate models of quantum gravity are the loop formulation and the spin network formulation. The reason is that the loop quantum gravity and the spin network quantum gravity, entail naturally the invariant polynomials of knots, links and tangles; and such polynomials are paradigmatic in the very recently developments of the TQC such as the AJL algorithm which was presented previously.

In the loop quantum gravity and in the spin network quantum gravity, the space-time is conceived as a very complex structure of. topological entanglement which is able to produce quantum entanglement and for then, quantum computation. Specifically the space-time is composed of braids, knots, links and tangles; and the amplitudes for the quantum gravitational process are determined by the invariant polynomials corresponding to these structures of topological entanglement. All this is precisely the physical background for the construction of quantum algorithms for gravitational quantum computers.

In quantum gravity a very special role is reserved for the structures of topological entanglement such as braid, knots, links and tangles. These structures are assumed constructed in the space-time in a such way that the results of measuring the quantum observables for the quantized gravitation field are given in terms of the invariants polynomials for knots, links and tangles [26,25,6,21]. It is possible to think about the quantum gravitational field as a quantum computer that is able to compute invariant polynomials such as Jones, HOMFLY, Kauffman and their generalizations including the Khovanov homology and the Khovanov-Rozansky homology.

To see how the link polynomials appear in the context of loop quantum gravity we use the following equations.

There are two pictures in loop quantum gravity [25]. The first one is called the Ahstekar connection representation and the second one is called the loop representation. A functional $\Phi(\mathbf{A})$ in the connection representation, being $\mathbf{A}$ an Ahstekar connection; and a functional $\Psi(\mathbf{L})$ in the loop representation, being $\mathbf{L}$ a link, are related by the loop transform whose kernel is the trace of the holonomy of the Ahstekar connection, it is to say the following Wilson loop for the circulation of $\mathbf{A}$ on $\mathbf{L}$ [25]

$$K[L, A] = \prod_{\lambda \in L} Tr(P \exp(\oint_{\lambda} A)),$$ (13)

and the following integral relation

$$\Psi(L) = \int K[L, A]\Phi(A)d\mu(A)$$ (14)

Where $d\mu(A)$ is a measure on the space of Ahstekar connections, that is assumed to be invariant under diffeomorphisms.

Now, a particularly important physical state for quantum gravity is the Kodama state which is defined in the connection representation in the following form [25]

$$\Phi(A) = N \exp(\frac{3}{8\pi\Lambda G}\int_{\Sigma} Y_{cs})$$ (15)

where Y_{CS} is the Chern-Simons action and it is defined as

$$Y_{CS} = \frac{1}{2}Tr(A \wedge dA + \frac{2}{3}A \wedge A \wedge A)$$ (16)

The loop transform of the Kodama state (15) according with (14) and (13), gives that

$$\Psi(L) = \int K[L, A]N \exp(\frac{3}{8\pi\Lambda G}\int_{\Sigma} Y_{CS})d\mu(A)$$ (17)

It is well known that (17) is justly the Witten partition function for the Chern-Simons theory with gauge group SU(2); and that the final result is $\Psi(L) = $ <L>(A,B,J), where <L>(A,B,J) is the Kauffman bracket defined by (1); it is to say that the Kodama state is the loop representation is the Kauffman bracket being the skein variable A certain function of Λ, where Λ is the cosmological constant [25]. All these show how the link polynomial naturally arise as non-perturbative physical states in loop quantum gravity.

In loop quantum gravity the relation between the Jones polynomial and the Kauffman bracket given by (3), is rewritten in the form [6]

$$K_{\Lambda}(L) = e^{-\Lambda w(L)}V_{\Lambda}(L)$$ (18)

Also it is well known that the action of the Hamitonian on the Kodama state (Kauffman bracket) is given by [6]

$$H_\Lambda K_\Lambda (L) = 0 \qquad (19)$$

This indicates that the Kodama state is an eigenstate of the Hamiltonian H_Λ associated with the eigen-energy zero.

Other two important restrictions which are very important for that it follows , read [6]

$$H_0 V_2 (L) = 0 \qquad (20)$$

$$H_0 V_3 (L) = 0 \qquad (21)$$

where $V_m(L)$ are the coefficients of the expansion of the Jones polynomial in terms of the cosmological constant. As is observed in (20) and (21), the coefficients of the Jones knot polynomial appear to be related to the solutions of the vacuum Hamiltonian constraint [6,25].

As is well known the quantum computing is based on one of the fundamental physical theories, namely the quantum theory. The other fundamental physical theory, the general relativity until now only has some few applications in computer science. In [27] an application of general relativity was given for the case of certain hard problems which are solved by a classic computer (Turing machine) when the computer is moved along of the geodesics of the curved space-time. In such computational model the gravitational field is not used as a computer. From other side in [28] the limitations of the quantum gravity on the quantum computers, were discussed but again the gravitational field is not considered as a computer by itself. More over in [29], the general relativity is used to compute justly non-computable problems including problems of higher degree of Turing non-computability. The strategy in such model of hypercomputation is to use a classical computer which is orbiting certain kind of rotating black hole. Again in this model the gravitational field is not used as a computer.

With all that is very interesting consider the possibility to have a computational model for which the gravitational field is the computer. This possibility can be explored both from the point of view of the general relativity as from the perspective of quantum gravity or more concretely from the optics of the loop quantum gravity.

Here we introduce a computational model which is named gravitational topological quantum computation and we expect that our model results relevant for computer science both from the side of complexity theory as from the side of computability theory.

As we said before the braid quantum computer of the anyonic topological quantum computation, is realized using anyon systems of condensed matter physics. Now, given the tangle is a general structure of topological entanglement it is possible to think about the tangle quantum computer . This tangle computer may be realized again on anyon systems but other possibility exists, namely the realization of the tangle computer in quantum gravity, given that quantum gravity contains tangles and Jones polynomials.

In this work we investigate only the gravitational topological tangle quantum computer which is realized using quantum gravitational fields. In the same way that the braid algebra is naturally associated with the anyonic systems, the tangle algebra is intimately associated with the quantum gravitational fields. In the anyonic TQC is expected that be possible to have quantum chips with a very large number of integrated quantum gates, simply with a certain braid whose strands are the universe lines for anyons. Similarly is possible to expect that the GTQC be able to produce such quantum chips, simply with a certain tangle which is the vacuum state of the quantum gravitational field.

In quantum computing there are various computational models such as the standard model with quantum circuits, the geometrical model with quantum gates derived from holonomy groups and the adiabatic model with its application of the adiabatic theorem of the quantum mechanics [30]. All these models are very well known in quantum computing and computer science and it is not necessary to describe them here.

In the present work we use the adiabatic method [30] and in the following subsection we present an adiabatic gravitational algorithm for the computation of the Kauffman bracket of a given link.

The AJL algorithm and its generalizations are proposed inside the domain of the anyonic topological quantum computing. As we said previously, in ATQC there are by at least four strategies to generate topological quantum algorithms for the computation of the Jones polynomial of a given link. The first is the original Freedman-Kitaev-Larsen-Wang using universal topological unitary functors and unitary representations of the braid groups. The second is the line Aharonov-Jones-Landau - Worcjan-Yard, using only algebra, it is to say, representation theory of the Temperley-Lieb algebras. The third is the methodology given by Rasseti-Sanardi using quantum spin networks as quantum cellular automata. And the fourth is the line Kauffman-Lomonaco-Zhang-Ge using yang-baxterization and the Kauffman's Temperley-Lieb recoupling theory. All these strategies are keeping at hand the anyon systems as the possible physical referents for a future implementation of the TQC.

But it is possible to think about other alternatives such as DNA computing and GTQC. In particular in GTQC is possible to formulate certain algorithm for the computation of the Kauffman bracket. The characteristic of our gravitational algorithm are:

It is based on the equations (13)-(21) which are showing the role of the Kauffman bracket in loop quantum gravity.

Our algorithm uses the well known adiabatic theorem of the quantum mechanics.

The physical computer for our algorithm is the quantum space-time.

The structure of our algorithm is the following:

Initialization is given using (20) and (21) for the case of an unknot, that is

$$H_0 V_2(\text{O}) = 0 \qquad (22)$$

$$H_0 V_3(\text{O}) = 0 \qquad (23)$$

according with

$$\left| I \right\rangle = a V_2(\text{O}) + b V_3(\text{O}) \qquad (24)$$

in such way that

$$H_0 \left| I \right\rangle = 0 \qquad (25)$$

The coding of the given link is realized using (13) for a certain Ahstekar connection in such way that (19) is satisfied. As we can to observe from (25) the initial state is an eigenstate of the Hamiltonian H_0 associated to the eigenvalue zero and as we can observe from (19) the Kauffman bracket is also an eigenstate of the Hamiltonian H_Λ corresponding to the eigenvalue zero.

The adiabatic interpolation between H_0 and H_Λ is given by the following Hamiltonian

$$H(t,T) = (1-\frac{t}{T})H_0 + (\frac{t}{T})H_\Lambda \tag{26}$$

where is immediately obvious that $H(0)=H_0$ and $H(T)=H_\Lambda$ being t the interpolation parameter and T the duration of execution of the algorithm.
Using the Hamiltonian (26) build the Schrodinger equation

$$i\hbar\frac{\partial}{\partial t}|\Psi(t)\rangle = H(t,T)|\Psi(t)\rangle \tag{27}$$

with the initial condition

$$|\Psi(0)\rangle = |I\rangle \tag{28}$$

where $|I>$ is given by (24)

Given a value of T, solve the Schrodinger equation (27) with (28) with the aim to obtain the state $|\Psi(T)>$.

If for the considered value of T, some component of $|\Psi(T)>$ has a probability more greater than 0.5, then the algorithm is finished and the dominant component of $|\Psi(T)>$ is justly the Kauffman bracket for given link L. This is due to the adiabatic theorem and the equation (19). But if $|\Psi(T)>$ has no a dominant component then a new more greater value of T is assigned and the algorithm is executed again.

This gravitational topological adiabatic quantum algorithm, constitutes an example of a quantum adiabatic speedup that relies on the topological order that it is inherent to the space-time in loop quantum gravity.

It is possible to think about the possible generalizations of our gravitational algorithm for the computation of other more stronger link invariants. For example given the actual developments in knot theory is very sensitive to think about the possibility to have topological quantum algorithms for invariant polynomials of knots, links and tangles, directly derived from very recently mathematical theories as the Khovanov cohomology [7]. In this point we declare that it is very interesting to try to exploit from the perspective of the quantum informatics in general and from the perspective of the GTQC in particular, the Khovanov cohomology for knots, links and tangles.

Specifically, we conjecture the existence of a quantum algorithm from the Khovanov comology This algorithm is initially written in a certain quantum high level language and for then it must to be compiled and translated to the quantum machine language corresponding to a quantum gravitational computer whose processors are tangles in the quantum space-time. In relation with this it is worthwhile to remark that in the paper [21] was given a construction of a family of tangle field theories related to the HOMFLY polynomial invariant of links. Such construction can be exploited from the perspective of the TQC in the sense that it is possible to have certain extended AJL algorithm inside the realm of the GTQC. It is due to that these tangle field theories produce unitary representations of the tangle group. From other side the calculations in tangle field theory are simplified because the basic equations in TFT include the skein relations for the HOMFLY polynomial.

More in detail, the mathematical algorithm invented by Khovanov has the following structure: given a knot or link or tangle, its generic plane projection is derived and associated with certain hypercube of smoothings and cobordisms. Then this hypercube is converted in a complex of homological algebra for which is associated the appropriate collection of bi-graded cohomology groups. The resulting cohomology generates the bi-graded Poincaré polynomial which gives the bi-graded Euler characteristic and this last is precisely the Jones polynomial.

This algorithm is originally for the categorification of the Jones polynomial but the Jones polynomial extends to a functor from the category of tangles to the category of vector spaces and for then the algorithm also gives the categorification of the topological invariants for tangles which are actually computed by the quantum gravitational process inside a tangled quantum space-time.

The extended functor associates a plane with n marked points with the tensorial power $V^{\otimes n}$, where V is the two-dimensional irreducible representation of the quantum group $U_q(sl_2)$. From the other side, the extended functor associates to an oriented tangle L with n bottom and m top endpoints , an operator $J(L) : V^{\otimes n} \rightarrow V^{\otimes m}$ which intertwines the $U_q(sl_2)$ action. In quantum informatics we have the following. The space V is interpreted as the space of a single qbit, the tensorial power $V^{\otimes n}$ is the space for n qbits, the tangle operator $J(L)$ is a quantum gate with an input of n qbits and an output of m qbits. And finally a qbit is certain quantum system whose dynamical symmetry is $U_q(sl_2)$ and for hence its quantum states are given by the two-dimensional irreducible representation of the quantum group $U_q(sl_2)$.

With all these, the Khovanov-BarNatan algorithm for the computation of tangle homology and its characteristic classes, has the same form that the previously described Khovanov algorithm for the Jones polynomial, the only difference is that now the Euler characteristic reproduces the Kauffman bracket extended to tangles.

Finally we mention the Khovanov-Rozansky algorithm which has the same structure than the previously described but with the difference that in this case the Euler characteristic reproduces the HOMFLY polynomial.

All these Khovanov algorithms are not yet quantum algorithms but we conjecture that the prominent role of tangles both in quantum gravity as in Khovanov homology is very plausible that a quantum algorithm for Khovanov homology can be based on GTQC.

From other side is very remarkable that recently, a relationship between Khovanov homology and topological strings, was discovered [31]. This discovery implies that is possible to use topological string theory in quantum computing as generalization of the usual TQFT, in this sense is coined the label Stringy Topological Quantum Computation (STQC) as a computational model which is a very closer relative to the GTQC.

4 Applications of GTQC in Computer Science

All models that were mentioned before, such as anyonic topological quantum computation, gravitational topological quantum computation, stringy topological quantum computation and DNA computing share in common that their computational powerful are directed to the solution of topology problems, specifically problems for braids, knots, links and tangles.

Many of these problems are hard problems and for hence all these alternative models and concretely the GTQC may be useful in computer science, specifically in the theory of algorithmic complexity.

From the other side other many problems in low dimension topology and knot theory are incomputable problems and again, the alternative models and particularly the GTQC may be useful in the theory of computability.

For hence in this section we discuss the possible applications of GTQC in computer science. Such applications can be in quantum complexity theory and in the computability theory.

It is well known that the simple Mathematica algorithm for computation of the Kauffman bracket of any knot and link, that was given in Table 1, has an asymptotic consumption of resources, such as time and space, which is exponential respect to the topological complexity of the knot input. In fact, the problem of computation of the Jones polynomial of a given link is a #P –hard problem [3,4,5]. More over, the generalized problems for the computation of the HOMFLY and Kauffman polynomials are also #P-hard [4,5]; but it is possible to conjecture that the problem of computation of the Khovanov cohomology of any link, is a problem with a more higher complexity than the #P-hardness, given the Khovanov polynomial is more powerful than the Jones polynomials [14].

From other side is very instructive to analyze the relationships between the link invariants, such as the Jones polynomial, HOMFLY polynomial, Kauffman polynomial and Khovanov homology; and the quantum computation. This analysis is a necessary preliminary step toward the understanding of the capabilities and limitations of the quantum computers in general and the topological quantum computers in particular such as the anyonic topological quantum computers, the gravitational topological quantum computers and the stringy topological quantum computers.

It is necessary to remark here that the AJL algorithm does not pretend to solve the #P-hard problem of the Jones polynomial, its intention is to solve the problem of approximating the Jones polynomial at certain values being such problem of approximation one that is BQP-hard. In this point is possible to conjecture that the approximated computations of the HOMFLY, Kauffman and Khovanov polynomials are again BQP-hard problems and for hence they demand all the computational power power of the quantum computing.

More in detail, the fact that the computation of the Jones polynomial is a #P-hard problem indicates that it is not possible to have a polynomial algorithm of any kind, classic or quantum , for the computation of the Jones polynomial, it is for the solution of a #P-hard problem. This affirmation can be sustained by the following arguments:

1. To spite that there is not a explicit proof of that NP is more hard than P, the majority of experts believe that it is true.
2. #P-hard is more hard than NP.
3. The problem of Jones polynomial is #P-hard

From these three arguments is derived that is very improbable that an algorithm exists in P for the computation of the Jones polynomial and similarly for the HOMFLY and Kauffman polynomials.

As we said before, actually is not known the algorithmic complexity of the problem of computation of the Khovanov homology for knots, links and tangles. We believe that this new problem is more hard than #P-hard. We believe that, because the Khovanov homology is a more stronger topological invariant that the Jones polynomial; and given the problem of Jones polynomial is #P-hard is natural to conjecture that the problem of Khovanov homology is very more difficult that #P-hard. More over, given that recently the Khovanov polynomial was considered as a particular case of the Lefschetz polynomial for link endo-cobordisms, it is possible to speculate that the computation of the Lefschetz polynomial is a problem which is more hard than the Khovanov polynomial which is believed more hard than #P-hard.

It is well known that there are many non-computable problems, it is to say non-computable by a Turing machine or DNA computer or a standard quantum computer (quantum circuits, geometrical quantum computer, anyonic topological quantum computer). A classical example of a non-computable problem is the Halting Problem which is equivalent to the Hilbert's tenth problem. This problem arises in number theory but it is possible to mention other non-computable problems which arise in algebra and topology, such as the word problem and the homeomorphism problem. It is necessary to say that the very low dimensional topology (d<4) appears as invulnerable with respect to the logical problems and for hence it is computable at full, given that the knot problem(to decide when an entanglement is really a knot and when two knots are equivalent), the word problem and the homeomorphism problem are soluble. But the Jones polynomial, the Kauffman polynomial and the Khovanov polynomial have not the capability of to solve the knot problem and it is an open problem the question that if the computable knot problem can be solved using some generalized kind of invariant polynomial.

Now, in the case of more higher dimensions (d>3) the knot problem, the word problem and the homeomorphism problem are non-computable. The word problem for groups refers to the question of to decide whether a given product of generators and their inverses is the identity and the homeomorphism problems refers to the question of to decide whether two topological spaces which are defined by a given pair of simplicial complexes are homeomorphic. The word problem and the homeomorphism problem have relationships with the Halting problem and the Hilbert's tenth problem and all these non-computable problems belong to the first grade of Turing non-computability, denoted as 0'.

Alright, in this point we make the claim that these non-computable problems can be computed using certain kind of hypercomputer which is based on GTQC. Specifically the knot problem and the homeomorphism problem can be solved in four dimensions using the formulation of the four dimensional quantum gravity as a topological quantum field theory whose observables are topological invariants. We conjecture that the homeomorphism problem can be translated in a problem of cobordism in quantum gravity and for hence it is possible to design an adiabatic quantum algorithm which has the capability of hyper-computing in the sense that such hyper-algorithm is able to solve a non-computable problem, namely the homeomorphism problem.

5 Conclusions

In this work certain model of gravitational topological quantum computation was presented. An algorithm for the computation of the Kauffman bracket was constructed. This algorithm is based on the topological adiabatic quantum computation and the loop quantum gravity. Some discussion about the possible applications of the GTQC in computer science were realized and some conjectures were formulated. These conjectures refer to the computational complexity of the Khovanov polynomial and its generalizations. Our conjecture says that such complexity is more higher that #P-hard but the complexity of the approximated problem remains BQP-hard.

In all case we believe that the tangles bi-categories and the Khovanov cohomology are very important in geometric topology and they can be relevant also in quantum gravity, topological quantum field theory, topological string theory, anyonic topological quantum computation, gravitational topological quantum computation and stringy topological quantum computation.

From other side, in the realm of the computability theory we made the claim that the GTQC can get the basis for the construction of a hyper-computer and the corresponding hyper-algorithm for the solution of non-computable problems such as the knot problem, the word problem and the homeomorphism problem.

From these perspectives we think that the new advances in knot theory and topological quantum computation are very important in computer science and for hence all these new alternative models of computation deserve an intensive investigation.

References

[1] Davis, M.: Computability and Unsolvability. McGraw-Hill Book Company, Inc. New York (1958)
[2] Chuang, I.L., Nielsen, M.A.: Quantum computation and quantum information. Cambridge University Press, Cambridge (2000)
[3] Freedman, M.H., Larsen, M., Wang, Z.: A modular Functor which is universal for quantum computation. Commun.Math.Phys. 227(3), 605–622 (2002)
[4] Aharonov, D., Jones, V., Landau, Z.: A polynomial quantum algorithm for approximating the Jones polynomial, quant-ph/0511096. In: STOC 06
[5] Wocjan, P., Yard, J.: The Jones polynomial: quantum algorithms and applicationsin quantum complexity theory, quant-ph/0603069
[6] Griego, J.: The Kauffman Bracket and the Jones polynomial in quantum gravity. Nucl. Phys. B 467, 332–354 (1996)
[7] Khovanov, M.: A categorification of the Jones polynomial. Duke Math. J. 101(3), 359–426 (2000)

[8] Witten, E.: Quantum field Theory and the Jones Polynomial. Commun. Math. Phys. 121, 351–399 (1989)

[9] Kauffman, L.H.: State models and the Jones polynomial. Topology 26, 395–407 (1987)

[10] Freedman, M., Kitaev, A., Larsen, M., Wang, Z.: Topological quantum computation. Mathematical challenges of the 21st century (Los Angeles, CA, 2000). Bull. Amer. Math. Soc (N.S.) 40(1), 31–38 (2003)

[11] Kauffman, L.H., Lomonaco, S.: q-Deformed Spin Networks, Knot Polynomials and Anyonic Topological Quantum Computation. quant-ph/0606114 v2

[12] Kauffman, L.H., Lomonaco, S.J.: Braiding Operators are Universal QuantumGates. New Journal of Physics 6(134), 1–39 (2004)

[13] Kauffman, L.H., Lomonaco, Jr., S.J.: Topological quantum computingand the Jones polynomial, quant-ph/0605004. In: SPIE Proceedings (2006)

[14] Bar-Natan, D.: On Khovanov's categorification of the Jones polynomial. Algebraic & Geometric Topology 2, 337–370 (2002)

[15] Jones, V.F.R: A polynomial invariant for knots via von Neumann algebras. Bull.Amer. Math. Soc. 12(1), 103–111 (1985)

[16] Kauffman, L.H., Lomonaco, Jr., S.J.: Entanglement Criteria - Quantum and Topological. In: Donkor, Pinch, Brandt. (eds.) Quantum information and computation-Spie Proceedings, 21-22 April, 2003, Orlando, FL, vol. 5105, pp. 51–58 (2003)

[17] Bigelow, S.: Braid Groups and Iwahori-Hecke Algebras, math-GT/0505064

[18] Artin, E.: Theory of braids. Annals of Mathematics 48, 101–126 (1947)

[19] Temperley, H., Lieb, E.: Relations between the 'percolation' and 'colouring' problem and other graph-theoretical problems associated with regular planar lattices: Some exact results for the 'percolation' problem. Proceedings of theRoyal Society of London A 322(1549), 251–280 (1971)

[20] Kauffman, L.H.: Temperley-Lieb Recoupling Theory and Invariants ofThree-Manifolds. In: Annals Studies, vol. 114, Princeton University Press, Princeton, NJ (1994)

[21] Baez, J.C.: Quantum Gravity and the Algebra of Tangles. Class. Quant.Grav. 10, 673–694 (1993)

[22] Zhang, Y., Kauffman, L.H., Ge, M.L.: Yang-Baxterizations, Universal QuantumGates and Hamiltonians. Quant. Inf. Proc. 4, 159–197 (2005)

[23] Garnerone, S., Marzuolli, A., Rasseti, M.: An efficient quantum algorithm for colored Jones polynomials, quant-ph/0606167

[24] Kitaev, A.: Anyons in an exactly solved model and beyond, arXiv.cond-mat/0506438 v1 17 (June 2005)

[25] Kauffman, L.H., Liko, T.: hep-th/0505069, Knot theory and a physical state of quantum gravity, Classical and Quantum Gravity, vol. 23, pp. R63 (2006)

[26] Rovelli, C., Smolin, L.: Spin networks and quantum gravity. Phys. Rev. D. 52, 5743–5759 (1995)

[27] Brun, T.A.: Computers with closed timelike curves can solve hard problems. Found. Phys. Lett. 16, 245–253 (2003)

[28] Srikanth, R.: The quantum measurement problem and physical reality:a computation theoretic perspective, quant-ph/0602114

[29] Etesi, G., Nemeti, I.: Non-Turing computations via Malament-Hogarth space-times. Int. J.Theor.Phys. 41, 341–370 (2002)

[30] Hamma, A., Lidar, D.A.: Topological Adiabatic Quantum Computation, quant-ph/0607145

[31] Gukov, S., Schwarz, A., Vafa, C.: Khovanov-Rozansky Homology and Topological Strings. Lett.Math.Phys. 74, 53–74 (2005)

Computation in
Sofic Quantum Dynamical Systems

Karoline Wiesner and James P. Crutchfield

Center for Computational Science & Engineering and Physics Department,
University of California Davis, One Shields Avenue, Davis, CA 95616
karoline@cse.ucdavis.edu, chaos@cse.ucdavis.edu

Abstract. We analyze how measured quantum dynamical systems store and process information, introducing sofic quantum dynamical systems. Using recently introduced information-theoretic measures for quantum processes, we quantify their information storage and processing in terms of *entropy rate* and *excess entropy*, giving closed-form expressions where possible. To illustrate the impact of measurement on information storage in quantum processes, we analyze two spin-1 sofic quantum systems that differ only in how they are measured.

1 Introduction

Symbolic dynamics originated as a method to study general dynamical systems when, nearly 100 years ago, Hadamard used infinite sequences of symbols to analyze the distribution of geodesics; see Ref. [1] and references therein. In the 1930's and 40's Hedlund and Morse coined the term *symbolic dynamics* [2,3]. In the 1940's Shannon used sequence spaces to describe information channels [4]. Subsequently, the techniques and ideas have found significant applications beyond dynamical systems, in data storage and transmission, as well as linear algebra [5].

On the other side of the same coin one finds finite-state automata that are used as codes for sequences. Sequences that can be coded this way define the *regular languages*. Dynamical systems can also be coded with finite-state automata. Transforming a dynamical system into a finite-state code is done using the tools from symbolic dynamics [5]. *Sofic systems* are the particular class of dynamical systems that are the analogs of regular languages in automata theory.

Yet another thread in the area of dynamical systems—the study of the quantum behavior of classically chaotic systems—gained much interest [6,7], most recently including the role of measurement. It turns out that measurement interaction leads to genuinely chaotic behavior in quantum systems, even far from the semi-classical limit [8]. Classical dynamical systems can be embedded in quantum dynamical systems as the special class of commutative dynamical systems, which allow for unambiguous assignment of joint probabilities to two observations [9].

An attempt to construct symbolic dynamics for quantum dynamical systems was made by Beck and Graudenz [10] and Alicki and Fannes [9]. Definitions

S.G. Akl et al.(Eds.): UC 2007, LNCS 4618, pp. 214–225, 2007.
© Springer-Verlag Berlin Heidelberg 2007

of entropy followed from this. However, no connection to quantum finite-state automata was established there.

Here we forge a link between quantum dynamical systems and quantum computation in general, and with quantum automata in particular, by extending concepts from symbolic dynamics to the quantum setting. The recently introduced computational model class of *quantum finite-state generators* provides the required link between quantum dynamical systems and the theory of automata and formal languages [11]. It gives access to an analysis of quantum dynamical systems in terms of symbolic dynamics. Here, we strengthen that link by studying explicit examples of quantum dynamical systems. We construct their quantum finite-state generators and establish their *sofic* nature. In addition we review tools that give an information-theoretic analysis for quantifying the information storage and processing of these systems. It turns out that both the sofic nature and information processing capacity depend on the way a quantum system is measured.

2 Quantum Finite-State Generators

To start, we recall the *quantum finite-state generators* (QFGs) defined in Ref. [11]. They consist of a finite set of *internal states* $Q = \{q_i : i = 1, \ldots, |Q|\}$. The *state vector* is an element of a $|Q|$-dimensional Hilbert space: $\langle\psi| \in \mathcal{H}$. At each time step a quantum generator outputs a symbol $s \in \mathcal{A}$ and updates its state vector.

The temporal dynamics is governed by a set of $|Q|$-dimensional *transition matrices* $\{T(s) = U \cdot P(s), s \in \mathcal{A}\}$, whose components are elements of the complex unit disk and where each is a product of a unitary matrix U and a projection operator $P(s)$. U is a $|Q|$-dimensional unitary *evolution operator* that governs the evolution of the state vector $\langle\psi|$. $\mathbf{P} = \{P(s) : s \in \mathcal{A}\}$ is a set of *projection operators*—$|Q|$-dimensional Hermitian matrices—that determines how the state vector is measured. The operators are mutually orthogonal and span the Hilbert space: $\sum_s P(s) = \mathbb{1}$.

Each output symbol s is identified with the measurement outcome and labels one of the system's eigenvalues. The projection operators determine how output symbols are generated from the internal, hidden unitary dynamics. They are the only way to observe a quantum process's current internal state.

A quantum generator operates as follows. U_{ij} gives the transition amplitude from internal state q_i to internal state q_j. Starting in state vector $\langle\psi_0|$ the generator updates its state by applying the unitary matrix U. Then the state vector is projected using $P(s)$ and renormalized. Finally, symbol $s \in \mathcal{A}$ is emitted. In other words, starting with state vector $\langle\psi_0|$, a single time-step yields $\langle\psi(s)| = \langle\psi_0| U \cdot P(s)$, with the observer receiving measurement outcome s.

2.1 Process Languages

The only physically consistent way to describe a quantum system under iterated observation is in terms of the observed sequence $\overleftrightarrow{S} \equiv \ldots S_{-2}S_{-1}S_0S_1\ldots$

of discrete random variables S_t. We consider the family of *word distributions*, $\{\Pr(s_{t+1}, \ldots, s_{t+L}) : s_t \in \mathcal{A}\}$, where $\Pr(s_t)$ denotes the probability that at time t the random variable S_t takes on the particular value $s_t \in \mathcal{A}$ and $\Pr(s_{t+1}, \ldots, s_{t+L})$ denotes the joint probability over sequences of L consecutive measurement outcomes. We assume that the distribution is stationary:

$$\Pr(S_{t+1}, \ldots, S_{t+L}) = \Pr(S_1, \ldots, S_L) . \tag{1}$$

We denote a block of L consecutive variables by $S^L \equiv S_1 \ldots S_L$ and the lowercase $s^L = s_1 s_2 \cdots s_L$ denotes a particular measurement sequence of length L. We use the term *quantum process* to refer to the joint distribution $\Pr(\overleftrightarrow{S})$ over the infinite chain of random variables. A quantum process, defined in this way, is the quantum analog of what Shannon referred to as an *information source* [12].

Such a quantum process can be described as a *stochastic language* $\mathcal{L}$, which is a *formal language* with a probability assigned to each word. A stochastic language's word distribution is normalized at each word length:

$$\sum_{\{s^L \in \mathcal{L}\}} \Pr(s^L) = 1 , L = 1, 2, 3, \ldots \tag{2}$$

with $0 \leq \Pr(s^L) \leq 1$ and the consistency condition $\Pr(s^L) \geq \Pr(s^L s)$.

A *process language* is a stochastic language that is *subword closed*: all sub-words of a word are in the language.

We can now determine word probabilities produced by a QFG. Starting the generator in $\langle \psi_0 |$, the probability of output symbol s is given by the state vector without renormalization:

$$\Pr(s) = \langle \psi(s) | \psi(s) \rangle . \tag{3}$$

While the probability of outcome s^L from a measurement sequence is

$$\Pr(s^L) = \langle \psi(s^L) | \psi(s^L) \rangle . \tag{4}$$

In [11] the authors established a hierarchy of process languages and the corresponding quantum and classical computation-theoretic models that can recognize and generate them.

2.2   Alternative Quantum Finite-State Machines

Said most prosaically, we view quantum generators as representations of the word distributions of quantum process languages. Despite similarities, this is a rather different emphasis than that used before. The first mention of *quantum automata* as an empirical description of physical properties was made by Albert in 1983 [13]. Albert's results were subsequently criticized by Peres for using an inadequate notion of measurement [14]. In a computation-theoretic context, quantum finite automata were introduced by several authors and in varying ways, but all as devices for recognizing word membership in a language. For the most widely discussed quantum automata, see Refs. [15,16,17]. Ref. [18] summarizes the different classes of languages which they can recognize. Quantum transducers were introduced by Freivalds and Winter [19]. Their definition, however, lacks a physical notion of measurement. We, then, introduced quantum finite-state machines, as a type of transducer, as the general object that can be reduced to the special cases of quantum recognizers and quantum generators of process languages [11].

3   Information Processing in a Spin-1 Dynamical System

We will now investigate concrete examples of quantum processes. Consider a spin-1 particle subject to a magnetic field which rotates the spin. The state evolution can be described by the following unitary matrix:

$$U = \begin{pmatrix} \frac{1}{\sqrt{2}} & \frac{1}{\sqrt{2}} & 0 \\ 0 & 0 & -1 \\ -\frac{1}{\sqrt{2}} & \frac{1}{\sqrt{2}} & 0 \end{pmatrix} . \tag{5}$$

Since all entries are real, U defines a rotation in $\mathbb{R}^3$ around the y-axis by angle $\frac{\pi}{4}$ followed by a rotation around the x-axis by an angle $\frac{\pi}{2}$.

Using a suitable representation of the spin operators J_i [20, p. 199]:

$$J_x = \begin{pmatrix} 0 & 0 & 0 \\ 0 & 0 & i \\ 0 & -i & 0 \end{pmatrix}, \ J_y = \begin{pmatrix} 0 & 0 & i \\ 0 & 0 & 0 \\ -i & 0 & 0 \end{pmatrix}, \ J_z = \begin{pmatrix} 0 & i & 0 \\ -i & 0 & 0 \\ 0 & 0 & 0 \end{pmatrix}, \tag{6}$$

the relation $P_i = 1 - J_i^2$ defines a one-to-one correspondence between the projector P_i and the square of the spin component along the i-axis. The resulting measurement answers the yes-no question, Is the square of the spin component along the i-axis zero?

Consider the observable J_y^2. Then the following projection operators together with U in Eq. (5) define a quantum finite-state generator:

$$P(0) = |010\rangle \langle 010|$$
$$\text{and} \ \ P(1) = |100\rangle \langle 100| + |001\rangle \langle 001| . \tag{7}$$

A graphical representation of the automaton is shown in Fig. 1. States A, B, and C correspond to the eigenstates $\langle 100|$, $\langle 010|$, and $\langle 001|$, respectively, of the observable J_y^2. The amplitudes attached to the transitions are given by the unitary matrix in Eq. (5). Since eigenstates A and C have degenerate eigenvalues the output symbol is the same when the machine enters either of these states.

The process language generated by this QFG is the so-called *Golden-Mean Process* language [1]. The word distribution is shown in Fig. 2. It is characterized

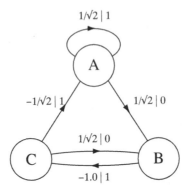

Fig. 1. The Golden Mean quantum generator. For details, see text.

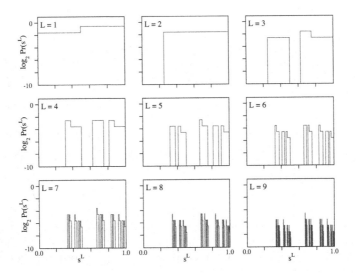

Fig. 2. Golden Mean process language: Word {00} has zero probability; all others have nonzero probability. The logarithm base 2 of the word probabilities is plotted versus the binary string s^L, represented as base-2 real number "$0.s^L$". To allow word probabilities to be compared at different lengths, the distribution is normalized on $[0, 1]$—that is, the probabilities are calculated as densities.

by the set of *irreducible forbidden words* $\mathcal{F} = \{00\}$: no consecutive zeros occur. In other words, for the spin-1 particle the spin is never measured twice in a row along the y-axis. This restriction—the dominant structure in the process—is a *short-range correlation* since the measurement outcome at time t only depends on the immediately preceding one at time $t - 1$. If the outcome is 0, the next outcome will be 1 with certainty. If the outcome is 1, the next measurement is maximally uncertain: outcomes 0 and 1 occur with equal probability.

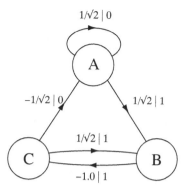

Fig. 3. The Even Process quantum generator

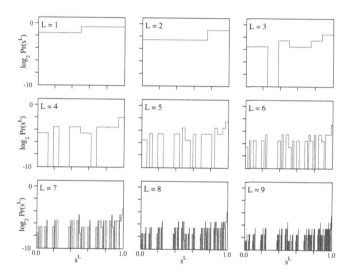

Fig. 4. Even Process language: Words $\{01^{2k-1}0\}, k = 1, 2, 3, \ldots$ have zero probability; all others have nonzero probability

Consider the same Hamiltonian, but now use instead the observable J_x^2. The corresponding projection operators are:

$$P(0) = |100\rangle \langle 100|$$
$$\text{and} \quad P(1) = |010\rangle \langle 010| + |001\rangle \langle 001| \ . \tag{8}$$

The QFG defined by U and these projection operators is shown in Fig. 3. States A, B, and C correspond to the eigenstates of the observable J_x^2. Since J_x^2 and J_y^2 commute they have the same eigenstates. The process language generated by this QFG is the so-called *Even Process* language [21,1]. The word distribution is shown in Fig. 4. It is defined by the infinite set of irreducible forbidden words $\mathcal{F} = \{01^{2k-1}0\}$, $k = 1, 2, 3, \ldots$. That is, if the spin component equals 0 along the x-axis it will be zero an even number of consecutive measurements before being observed to be nonzero. This is a type of *infinite correlation*: For a possibly infinite number of time steps the system tracks the evenness or oddness of the number of consecutive measurements of "spin component equals 0 along the x-axis".

Note that changing the measurement, specifically choosing J_z^2 as the observable, yields a QFG that generates *Golden Mean* process language again.

The two processes produced by these quantum dynamical systems are well known in the context of symbolic dynamics [5]—a connection we will return to shortly. Let us first turn to another important property of finite-state machines and explore its role in computational capacity and dynamics.

4   Determinism

The label *determinism* is used in a variety of senses, some of which are seemingly contradictory. Here, we adopt the notion, familiar from automata theory [22], which differs from that in physics, say, of non-stochasticity. One calls a finite-state machine (classical or quantum) *deterministic* whenever the transition from one state to the next is uniquely determined by the output symbol (or input symbol for recognizers). It is important to realize that a *deterministic* finite-state machine can still behave stochastically—stochasticity here referring to the positive probability of generating symbols. Once the symbol is determined, though, the transition taken by the machine to the next state is unique. Thus, what is called a *stochastic process* in dynamical systems theory can be described by a deterministic finite-state generator without contradiction.

We can easily check the two quantum finite-state machines in Figs. 1 and 3 for determinism by inspecting each state and its outgoing transitions. One quickly sees that both generators are deterministic. In contrast, the third QFG mentioned above, defined by U in Eq. (5) and J_z^2, is nondeterministic.

Determinism is a desirable property for various reasons. One is the simplicity of the mapping between observed symbols and internal states. Once the observer synchronizes to the internal state dynamics, the output symbols map one-to-one *onto* the internal states. In general, though, the observed symbol sequences do not track the internal state dynamics (orbit) directly. (This brings one to the topic of hidden Markov chains [23].)

For optimal prediction, however, access to the internal state dynamics is key. Thus, when one has a deterministic model, the observed sequences reveal the internal dynamics. Once they are known and one is synchronized, the process becomes optimally predictable. A final, related reason why determinism is desirable is that closed-form expressions can be given for various information processing measures, as we will discuss in Sec. 6.

5   Sofic Systems

In symbolic dynamics, sofic systems are used as tractable representations with which to analyze continuous-state dynamical systems [5,1]. Let the alphabet $\mathcal{A}$ together with an $n \times n$ adjacency matrix (with entries 0 or 1) define a directed graph $G = (V, E)$ with V the set of vertices and E the set of edges. Let X be the set of all infinite admissible sequences of edges, where *admissible* means that the sequence corresponds to a path through the graph. Let T be the shift operator on this sequence; it plays the role of the time-evolution operator of the dynamical system. A *sofic system* is then defined as the pair (X, T) [24]. The Golden Mean and the Even process are standard examples of *sofic systems*. The Even system, in particular, was introduced by Hirsch et al in the 1970s [21].

Whenever the rule set for admissible sequences is finite one speaks of a *subshift of finite type*. The Golden Mean process is a subshift of finite type. Words in the language are defined by the finite (single) rule of not containing the subword 00. The Even Process, on the other hand, is not of finite type, since the number of rules is infinite: The forbidden words $\{01^{2k+1}0\}$ cannot be reduced to a finite set. As we noted, the rule set, which determines allowable words, implies the process has a kind of infinite memory. One refers, in this case, to a *strictly sofic* system.

The spin-1 example above appears to be the first time a strictly sofic system has been identified in quantum dynamics. This ties quantum dynamics to languages and quantum automata theory in a way similar to that found in classical dynamical systems theory. In the latter setting, words in the sequences generated by sofic systems correspond to *regular languages*—languages recognized by some finite-state machine. We now have a similar construction for quantum dynamics. For any (finite-dimensional) quantum dynamical system under observation we can construct a QFG, using a unitary operator and a set of projection operators. The language it generates can then be analyzed in terms of the rule set of admissible sequences. One interesting open problem becomes the question whether the words produced by sofic quantum dynamical systems correspond to the regular languages. An indication that this is not so is given by the fact that finite-state quantum recognizers can accept nonregular process languages [11].

6   Information-Theoretic Analysis

The process languages generated by the spin-1 particle under a particular observation scheme can be analyzed using well known information-theoretic quantities such as *Shannon block entropy* and *entropy rate* [12] and others introduced in Ref. [25]. Here, we will limit ourselves to the *excess entropy*. The applicability of this analysis to quantum dynamical systems has been shown in Ref. [26], where closed-form expressions are given for some of these quantities when the generator is known.

We can use the observed behavior, as reflected in the word distribution, to come to a number of conclusions about how a quantum process generates randomness and stores and transforms historical information. The *Shannon entropy* of length-L sequences is defined

$$H(L) \equiv - \sum_{s^L \in \mathcal{A}^L} \Pr(s^L) \log_2 \Pr(s^L) \ . \tag{9}$$

It measures the average surprise in observing the "event" s^L. Ref. [25] showed that a stochastic process's informational properties can be derived systematically by taking derivatives and then integrals of $H(L)$, as a function of L. For example, the *source entropy rate* h_μ is the rate of increase with respect to L of the Shannon entropy in the large-L limit:

$$h_\mu \equiv \lim_{L \to \infty} [H(L) - H(L-1)] \ , \tag{10}$$

where the units are *bits/measurement* [12].

Ref. [26] showed that the entropy rate of a quantum process can be calculated directly from its QFG, when the latter is deterministic. A closed-form expression for the entropy rate in this case is given by:

$$h_\mu = -|Q|^{-1} \sum_{i=0}^{|Q|-1} \sum_{j=0}^{|Q|-1} |U_{ij}|^2 \log_2 |U_{ij}|^2 \ , \tag{11}$$

The entropy rate h_μ quantifies the irreducible randomness in processes: the randomness that remains after the correlations and structures in longer and longer sequences are taken into account.

The latter, in turn, is measured by a complementary quantity. The amount $I(\overleftarrow{S}; \overrightarrow{S})$ of mutual information [12] shared between a process's past $\overleftarrow{S}$ and its future $\overrightarrow{S}$ is given by the *excess entropy* **E** [25]. It is the subextensive part of $H(L)$:

$$\mathbf{E} = \lim_{L \to \infty} [H(L) - h_\mu L] \ . \tag{12}$$

Note that the units here are *bits*.

Ref. [25] gives a closed-form expression for **E** for *order-R* Markov processes—those in which the measurement symbol probabilities depend only on the previous $R-1$ symbols. In this case, Eq. (12) reduces to:

$$E = H(R) - R \cdot h_\mu \ , \tag{13}$$

where $H(R)$ is a sum over $|\mathcal{A}|^R$ terms. Given that the quantum generator is deterministic we can simply employ the above formula for h_μ and compute the block entropy at length R to obtain the excess entropy for the order-R quantum process.

Ref. [26] computes these entropy measures for various example systems, including the spin-1 particle. The results are summarized in Table 1. The value for the excess entropy of the Golden Mean process obtained by using Eq. (13) agrees with the value obtained from simulation data, shown in Table 1. The

Table 1. Information storage and generation for example quantum processes: entropy rate h_μ and excess entropy **E**

Quantum Dynamical System Spin-1 Particle		
Observable	J_y^2	J_x^2
h_μ [bits/measurement]	0.666	0.666
E [bits]	0.252	0.902

entropy $h_\mu = 2/3$ bits per measurement for both processes, and thus they have the same amount of irreducible randomness. The excess entropy, though, differs markedly. The Golden Mean process (J_y^2 measured) stores, on average, $\mathbf{E} \approx 0.25$ bits at any given time step. The Even Process (J_x^2 measured) stores, on average, $\mathbf{E} \approx 0.90$ bits, which reflects its longer memory of previous measurements.

7   Conclusion

We have shown that quantum dynamical systems store information in their dynamics. The information is accessed via measurement. Closer inspection would suggest even that information is *created* through measurement. The key conclusion is that, since both processes are represented by a 3-state QFG constructed from the same internal quantum dynamics, it is the means of observation alone that affects the amount of memory. This was illustrated with the particular examples of the spin-1 particle in a magnetic field. Depending on the choice of observable the spin-1 particle generates different process languages. We showed that these could be analyzed in terms of the block entropy—a measure of uncertainty, the entropy rate—a measure of irreducible randomness, and the excess entropy—a measure of structure. Knowing the (deterministic) QFG representation, these quantities can be calculated in closed form.

We established a connection between quantum automata theory and quantum dynamics, similar to the way *symbolic dynamics* connects classical dynamics and automata. By considering the output sequence of a repeatedly measured quantum system as a shift system we found quantum processes that are sofic systems. Taking one quantum system and observing it in one way yields a subshift of finite type. Observing it in a different way yields a (strictly sofic) subshift of infinite type. Consequently, not only the amount of memory but also the soficity of a quantum process depend on the means of observation.

This can be compared to the fact that, classically the Golden Mean and the Even sofic systems can be transformed into each other by a two-block map. The adjacency matrix of the graphs is the same. A similar situation arises here. The unitary matrix, which is the corresponding adjacency matrix of the quantum graph, is the same for both processes.

The preceding attempted to forge a link between quantum dynamical systems and quantum computation by extending concepts from symbolic dynamics to the quantum setting. We believe the results suggest further study of the properties of quantum finite-state generators and the processes they generate is necessary and will shed light on a number of questions in quantum information processing. One open technical question is whether sofic quantum systems are the closure of quantum subshifts of finite-type, as they are for classical systems [24]. There are indications that this is not so. For example, as we noted, quantum finite-state recognizers can recognize nonregular process languages [11].

References

1. Kitchens, B.P.: Symbolic dynamics: one-sides, two-sided, and countable state Markov shifts. Springer, Heidelberg (1998)
2. Hedlund, G.H., Morse, M.: Symbolic dynamics i. Amer. J. Math. 60, 815–866 (1938)
3. Hedlund, G.H., Morse, M.: Symbolic dynamics ii. Amer. J. Math. 62, 1–42 (1940)
4. Shannon, C.E., Weaver, W.: The Mathematical Theory of Communication. University of Illinois Press, Champaign-Urbana (1962)
5. Lind, D., Marcus, B.: An introduction to symbolic dynamics and coding. Cambridge University Press, Cambridge (1995)
6. Gutzwiller, M.C.: Chaos in Classical and Quantum Mechanics. Springer, Heidelberg (1990)
7. Reichl, L.E.: The transition to chaos: Conservative classical systems and quantum manifestations. Springer, New York (2004)
8. Habib, S., Jacobs, K., Shizume, K.: Emergence of chaos in quantum systems far from the classical limit. Phys. Rev. Lett. 96, 10403–10406 (2006)
9. Alicki, R., Fannes, M.: Quantum dynamical systems. Oxford University Press, Oxford (2001)
10. Beck, C., Graudenz, D.: Symbolic dynamics of successive quantum-mechanical measurements. Phys. Rev. A 46(10), 6265–6276 (1992)
11. Wiesner, K., Crutchfield, J.P.: Computation in finitary quantum processes. e-print arxiv/quant-ph/0608206 (submitted 2006)
12. Cover, T., Thomas, J.: Elements of Information Theory. Wiley-Interscience, Chichester (1991)
13. Albert, D.Z.: On quantum-mechanical automata. Physics Letters 98A, 249–251 (1983)
14. Peres, A.: On quantum-mechanical automata. Physics Letters 101A, 249–250 (1984)
15. Moore, C., Crutchfield, J.P.: Quantum automata and quantum grammars. Theor. Comp. Sci. 237, 275–306 (2000)
16. Kondacs, A., Watrous, J.: On the power of quantum finite state automata. In: 38th IEEE Conference on Foundations of Computer Science, pp. 66–75. IEEE Computer Society Press, Los Alamitos (1997)
17. Aharonov, D., Kitaev, A., Nisan, N.: Quantum circiuts with mixed states. In: 30th Annual ACM Symposium on the Theory of Computing, pp. 20–30. ACM Press, New York (1998)
18. Ambainis, A., Watrous, J.: Two-way finite automata with quantum and classical states. Theoretical Computer Science 287, 299–311 (2002)

19. Freivalds, R., Winter, A.: Quantum finite state transducers. Lect. Notes Comp. Sci. 2234, 233–242 (2001)
20. Peres, A.: Quantum theory: concepts and methods. Kluwer Academic Publishers, Dordrecht (1993)
21. Hirsch, M., Palis, J., Pugh, C., Shu, M.: Neighborhoods of hyperbolic sets. Inventiones Math. 9, 121–134 (1970)
22. Hopcroft, J.E., Motwani, R., Ullman, J.D.: Introduction to Automata Theory, Languages, and Computation. Addison-Wesley, Reading (2001)
23. Bishop, C.M.: Pattern recognition and machine learning. Springer Verlag, Singapore (2006)
24. Weiss, B.: Subshifts of finite type and sofic systems. Monatshefte für Mathematik 77, 462–474 (1973)
25. Crutchfield, J.P., Feldman, D.P.: Regularities unseen, randomness observed: Levels of entropy convergence. Chaos 13, 25–54 (2003)
26. Crutchfield, J.P., Wiesner, K.: Intrinsic quantum computation. e-print arxiv/quant-ph/0611202 (submitted 2006)

Bond Computing Systems: A Biologically Inspired and High-Level Dynamics Model for Pervasive Computing*

Linmin Yang[1], Zhe Dang[1], and Oscar H. Ibarra[2]

[1] School of Electrical Engineering and Computer Science
Washington State University
Pullman, WA 99164, USA
[2] Department of Computer Science
University of California
Santa Barbara, CA 93106, USA

Abstract. Targeting at modeling the high-level dynamics of pervasive computing systems, we introduce Bond Computing Systems (BCS) consisting of objects, bonds and rules. Objects are typed but addressless representations of physical or logical (computing and communicating) entities. Bonds are typed multisets of objects. In a BCS, a configuration is specified by a multiset of bonds, called a collection. Rules specifies how a collection evolves to a new one. A BCS is a variation of a P system introduced by Gheorghe Paun where, roughly, there is no maximal parallelism but with typed and unbounded number of membranes, and hence, our model is also biologically inspired. In this paper, we focus on regular bond computing systems (RBCS), where bond types are regular, and study their computation power and verification problems. Among other results, we show that the computing power of RBCS lies between LBA (linearly bounded automata) and LBC (a form of bounded multicounter machines) and hence, the regular bond-type reachability problem (given an RBCS, whether there is some initial collection that can reach some collection containing a bond of a given regular type) is undecidable. We also study some restricted models (namely, B-boundedness and 1-transfer) of RBCS where the reachability problem becomes decidable.

1   Introduction

As inexpensive computing devices (e.g., sensors, celluar phones, PDAs, etc.) become ubiquitous in our daily life, pervasive computing [14], a proposal of building distributed software systems from (a massive number of) such devices that are pervasively hidden in the environment, is no longer a dream. An example of pervasive computing systems is Smart Home [7], which aims to provide a set of

---

* The work by Linmin Yang and Zhe Dang was supported in part by NSF Grant CCF-0430531. The work by Oscar H. Ibarra was supported in part by NSF Grant CCF-0430945 and CCF-0524136.

S.G. Akl et al.(Eds.): UC 2007, LNCS 4618, pp. 226–241, 2007.

intelligent home appliances that make our life more comfortable and efficient. In Smart Home, a classic scenario is that as soon as a person goes back home from the office, the meal is cooked and the air conditioner is turned on, etc. Another example is a health monitoring network for, e.g., assisting the aged people [1]. In the system, a typical scenario (which is also used throughout the paper) is that when one gets seriously sick, a helicopter will, automatically, come and carry him/her to a clinic.

However, how to design a complex pervasive computing application remains a grand challenge. A way to respond to the challenge is to better understand, both fundamentally and mathematically, some key views of pervasive computing. Following this way, formal computation models reflecting those views are highly desirable and, additionally, can provide theoretical guidelines for further developing formal analysis (such as verification and testing) techniques for pervasive computing systems.

At a high-level, there are at least two views in modeling a pervasive computing system. From the local view, one needs to formally model how a component or an object interacts with the environment with the concept, which is specific to pervasive computing, of context-awareness [12] in mind using notions like Context UNITY [10] and Context-Aware Calculus [15]. From the global view, one needs to formally model how a group of objects evolve and interact with others, while abstracting the lower-level communications away. An example of such a view is the concept of *communities* in the design frame work PICO [13,4], where objects are (dynamically) grouped and collaborated to perform application-specific services. Inspired by the communities concept in PICO, we introduce a formal computation model in this paper, called *bond computing systems* (BCS), to reflect the global view for pervasive computing.

The basic building blocks of a BCS are *objects*, which are logical representations of physical (computing and communicating) entities. An object is typed but not addressed (i.e., without an individual identifier; we leave addressing to the lower-level (network) implementation). For instance, when a person is seriously sick, we care about *whether* a helicopter will come instead of *which* helicopter will come, under the simplest assumption that all helicopters share the same type *helicopter*.

In a BCS, objects are grouped in a *bond* to describe how objects are associated. For instance, a helicopter carrying a patient can be considered a bond (of two objects). Later, when the helicopter picks up one more patient, the bond then contains three objects. Since objects of the same type share the same name, a bond is essentially a multiset of (typed) objects. A bond itself also has a type to restrict how many and what kinds of objects that can be in the bond. In our system, bonds do not overlap; i.e., one object cannot appear in more than one bond at the same time (we leave overlapping and nested bonds in future work).

In a BCS, a configuration is specified by a multiset of bonds, called a *collection*. The BCS has a number of *rules*, each of which specifies how a collection evolves to a new one. Basically, a rule consists of a number of atomic rules that are fired in parallel, and each atomic rule specifies how objects transfer from a typed bond to another.

Our bond computing systems are a variation of P systems introduced by Gheorghe Paun [8]. A P system is an unconventional computing model motivated from natural phenomena of cell evolutions and chemical reactions. It abstracts from the way living cells process chemical compounds in their compartmental structures. Thus, regions defined by a membrane structure contain objects that evolve according to given rules. The objects can be described by symbols or by strings of symbols, in such a way that multisets of objects are placed in regions of the membrane structure. The membranes themselves are organized as a Venn diagram or a tree structure where one membrane may contain other membranes. By using the rules in a nondeterministic, maximally parallel manner, transitions between the system configurations can be obtained. A sequence of transitions shows how the system is evolving.

Our bond computing systems can be considered as P systems and hence are also biologically inspired. The important features of a BCS, which are tailored for pervasive computing, are the following:

1. The objects in a BCS can not evolve into other objects (e.g., a *helicopter* object can not evolve into a *doctor* object in a BCS). This is because in pervasive computing systems, objects while offering services, mostly maintain their internal structures.
2. The number of bonds in a BCS is unbounded. This makes it flexible to specify a pervasive application with many instances of a community.
3. Each rule in a BCS is global, since the BCS is intended to model the dynamics of a pervasive computing system as a whole rather than a specific bond. If applicable, every two bonds can communicate with each other via a certain rule. In tissue P systems [6,9], membranes are connected as a directed graph, and one membrane can also communicate with others if there are rules between them, without the constraint of tree-like structures. However, in tissue P systems, every rule can only connect two designated membranes, and in BCS, bonds are addressless and hence one rule can be applied to multiple pairs of bonds. Furthermore, the number of membranes in a tissue P system is bounded.
4. There is no maximal parallelism in a BCS. Maximal parallelism, like in P systems, is extremely powerful [8,3]. However, considering a pervasive computing model that will eventually be implemented over a network, the cost of implementing the maximal parallelism (which requires a global lock) is unlikely realistic, and is almost impossible in unreliable networks [5]. Hence, rules in a BCS fire asynchronously, while inside a rule, some local parallelism is allowed. Notice that P systems without maximal parallelism have been studied, e.g., in [2], where the number of membranes is fixed, while in BCS, the number of nonempty bonds could be unbounded.

In this paper, we focus on *regular bond computing systems* (RBCS), where bond types are regular, and study their computation power and verification problems. Among other results, we show that the computing power of RBCS lies between LBA (linearly bounded automata) and LBC (a form of bounded multicounter machines) and hence, the following *bond-type reachability problem* is undecidable: whether, for a given RBCS and a regular bond type L, there is some initial collection that can reach some collection containing a bond of type L. Given this undecidability result, we then study two forms of restrictions that can be respectively applied on RBCS such that the bond-type reachability problem becomes decidable. The first form of restricted RBCS is called B-bounded, where each bond contains at most B objects. In this case, the bond-type reachability problem is decidable by showing B-bounded RBCS can be simulated by vector addition systems (i.e., Petri nets). The second form of restricted RBCS is called 1-transfer, where each rule contains only one atomic rule and transfers only one type of objects. In this case, the bond-type reachability problem is also decidable by introducing a symbolic representation of a set of reachable collections.

The rest of the paper is organized as follows. In Section 2, we give the definition of bond computing systems along with an example. In Section 3, we study the computing power of regular bond computing systems and show that the bond-type reachability problem is undecidable in general for regular bond computing systems. In Section 4, we study further restrictions on regular bond computing systems such that the reachability problem becomes decidable. We conclude the paper in Section 5.

2   Definitions

Let $\Sigma = \{a_1, ..., a_k\}$ be an alphabet of symbols, where $k \geq 1$. An instance of a symbol is called an *object*. A *bond* w is a finite multiset over Σ, which is represented by a string in $a_1^*...a_k^*$. For example, the bond $w = a_2^3 a_4^2 a_7^5$ is the multiset consisting of three a_2's, two a_4's, and five a_7's. The empty (null) multiset is denoted by the null string λ.

Throughout this paper, a language L over Σ will mean a subset of $a_1^*...a_k^*$ (representing a set of multisets over Σ). $L_\emptyset$ will denote the language containing only λ. Hence, a *bond type*, that is a (possibly infinite) set of finite multisets, is simply a language L. Let $\mathcal{L} = \{L_\emptyset, L_1, ..., L_l\}$ be $l + 1$ languages (i.e., bond types), where $l \geq k$, and for $1 \leq i \leq k$, $L_i = \{a_i\}$ (i.e., L_i is the language containing only the symbol a_i). We will usually denote a regular language by a regular expression, e.g., a_i instead of $\{a_i\}$.

A *bond collection* $\mathcal{C}$ is an infinite multiset of bonds, where there are only finitely many nonempty bonds. A bond in the collection may belong to one or more bond types in $\mathcal{L}$ or may not belong to any bond type in $\mathcal{L}$ at all.

Example 1. Consider the alphabet $\Sigma = \{p, h, c, d\}$, where p, h, c and d represent *patient, helicopter, clinic* and *doctor*, respectively. We also have bond types $\mathcal{L} = \{L_\emptyset, L_1, L_2, L_3, L_4, L_5, L_6, L_7, L_8, L_9, L_{10}\}$, where each L_i is specified by a regular expression as follows:

- By definition, $L_1 = p$, $L_2 = h$, $L_3 = c$ and $L_4 = d$.

- $L_5 = p^*h$ is to specify a "helicopter-patients" bond (i.e., a helicopter carrying 0 or more patients).
- $L_6 = p^{\leq 8}c(d+dd)$ is to specify a "clinic-patients-doctors" bond (i.e., a clinic housing at most 8 patients and with 1 or 2 doctors).
- $L_7 = p^8cd^2$ is to specify a "clinic-patients-doctors" bond that has 8 patients and 2 doctors exactly.
- $L_8 = p^8cd^3$ is to specify a "clinic-patients-doctors" bond that has 8 patients and 3 doctors exactly.
- $L_9 = cd$ is used to specify a "clinic-patients-doctors" bond but with one doctor and no patient.
- $L_{10} = p^{>8}cd^{>2}$ is to specify a "clinic-patients-doctors" bond but with more than 8 patients and more than 2 doctors.

A bond collection $\mathcal{C}$ specifies a scenario of how patients, helicopters and clinics are bonded together. For instance, let $\mathcal{C}$ be a multiset of bonds

$$\{p, p, h, d, d, p^2h, ph, p^4cd, p^8cd^2\}$$

(we omit all the empty bonds λ when writing $\mathcal{C}$). This $\mathcal{C}$ indicates that we have totally 17 patients, 3 helicopters, 2 clinics, and 5 doctors, where one helicopter carries 2 patients, one helicopter carries 1 patient, and the remaining helicopter carries none. Also in the collection, one clinic has 4 patients and 1 doctor (while the other one has 8 patients and two doctors). This collection is depicted in the left hand side of Figure 1. Notice that a bond does not neccessarily mean a "physical bond" (like a patient boarding a helicopter). For instance, the patient and the insurance company with which the patient is subscribed are "logically" (instead of "physically") bonded.   □

A *rule* specifies how objects are transferred between bonds with certain types. Formally, a *rule* R consists of a finite number of *atomic rules* r, each of which is in the following form:

$$(L)_{\text{left}_r} \xrightarrow{\alpha_r} (L')_{\text{right}_r}, \tag{1}$$

where left_r and right_r are numbers, called the left index and the right index, respectively, L and L' are bond types in $\mathcal{L}$, and string α_r represents the finite multiset of objects that will be transferred.

The semantics of the rule R, when applied on a collection $\mathcal{C}$ of bonds, is as follows. We use I to denote the set of all the left and right indices that appear in the atomic rules r in R. Without loss of generality, let $I = \{1, \cdots, m\}$ for some m.

For each $i \in I$, let w_i be a distinct bond in the collection. We say that the rule R *modifies* $\langle w_1, \cdots, w_m \rangle$ into $\langle w'_1, \cdots, w'_m \rangle$, when all of the following conditions are satisfied:

- For each appearance $(L)_i$, with $L \in \mathcal{L}$ and $i \in I$, in the rule R, w_i must be of bond type L.
- For each $i \in I$, we use LEFT_i to denote the set of all atomic rules r in R such that the left index left_r of r equals i. Then, each w_i must contain all

the objects that ought to be transferred, i.e., the multiset union of multisets α_r in each atomic rule r that is in LEFT$_i$.

- The bonds $\langle w'_1, \cdots, w'_m \rangle$ are the result of $\langle w_1, \cdots, w_m \rangle$ after, *in parallel* for all atomic rules r in R, transferring objects in multiset α_r from bond w_{left_r} to bond w_{right_r}. That is, $\langle w'_1, \cdots, w'_m \rangle$ are the outcome of the following pseudo-code:

```
For each r ∈ R
    w_left_r := w_left_r − α_r;    // '−' denotes multiset difference
For each r ∈ R
    w_right_r := w_right_r + α_r;  // '+' denotes multiset union
For each i ∈ I
    w'_i := w_i.
```

When there are some $\langle w_1, \cdots, w_m \rangle$ in the collection $\mathcal{C}$ such that R modifies $\langle w_1, \cdots, w_m \rangle$ into some $\langle w'_1, \cdots, w'_m \rangle$, we say that R is *enabled* and, furthermore, R *modifies* the collection $\mathcal{C}$ into collection $\mathcal{C}'$, written

$$\mathcal{C} \xrightarrow{R} \mathcal{C}'$$

where $\mathcal{C}'$ is the result of replacing $w_1, \cdots, w_m$ in $\mathcal{C}$ with $w'_1, \cdots, w'_m$, respectively.

Example 2. Let bond collection $\mathcal{C}$ be the one defined in Example 1. That is, $\mathcal{C}$ is the multiset of bonds $\{p, p, h, d, d, p^2h, ph, p^4cd, p^8cd^2\}$. Recall that p, h, c and d represent patient, helicopter, clinic and doctor, respectively. Consider a rule, called TRANSFER_CALL-IN, that describes the scenario that a helicopter transfers a patient to a clinic that already houses 8 patients but with only two doctors on duty. In parallel with the transferring, an on-call doctor (who is not bonded with a clinic yet) is called in and bonded with the clinic to treat the patient. The rule TRANSFER_CALL-IN consists of the following two atomic rules:

TRANSFER: $(p^*h)_1 \xrightarrow{p} (p^8cd^2)_2$

CALL-IN: $(d)_3 \xrightarrow{d} (p^8cd^2)_2$

In order to describe how the rule is fired, by definition, we (nondeterministically) choose three bonds from the collection $\mathcal{C}$: $\langle w_1, w_2, w_3 \rangle = \langle p^2h, p^8cd^2, d \rangle$. Notice that bond w_1 corresponds to the term $(p^*h)_1$ in the atomic rule TRANSFER; bond w_2 corresponds to the term $(p^8cd^2)_2$ in the atomic rule TRANSFER and in the atomic rule CALL-IN; and bond w_3 corresponds to the term $(d)_3$ in the atomic rule CALL-IN. (Note that the subscript i of w_i equals the subscript of the term correspondent to the bond w_i.) One can easily check that these three bonds are respectively of the bond types specified in the terms. Firing the rule on the chosen $\langle w_1, w_2, w_3 \rangle$ will make a p transferred from bond w_1 to w_2 and make a d transferred from bond w_3 to w_2, in parallel. That is, $\langle w_1, w_2, w_3 \rangle$ is modified into $\langle w'_1, w'_2, w'_3 \rangle = \langle ph, p^9cd^3, \lambda \rangle$. Replacing the previously chosen $\langle w_1, w_2, w_3 \rangle$ in $\mathcal{C}$ with the $\langle w'_1, w'_2, w'_3 \rangle$, we get $\mathcal{C}' = \{p, p, h, d, ph, ph, p^4cd, p^9cd^3\}$ (again, we omit the empty bonds λ). Hence, we have $\mathcal{C} \xrightarrow{R} \mathcal{C}'$ (where R is the rule

TRANSFER_CALL-IN) as depicted in Figure 1. Notice also that, depending on the choice of $\langle w_1, w_2, w_3 \rangle$, the resulting collection $\mathcal{C}'$ may not be unique; e.g., one can choose $\langle w_1, w_2, w_3 \rangle = \langle ph, p^8cd^2, d \rangle$ from $\mathcal{C}$ and, subsequently, $\mathcal{C}' = \{p, p, h, d, p^2h, h, p^4cd, p^9cd^3\}$. Finally, we shall point out that if we change the rule TRANSFER_CALL-IN into the following form:

TRANSFER: $(p^*h)_1 \overset{ppp}{\mapsto} (p^8cd^2)_2$

CALL-IN: $(d)_3 \overset{d}{\rightarrow} (p^8cd^2)_2$

then this rule is not enabled under the configuration $\mathcal{C}$. (Since, now, the rule, when fired, *must* transfer 3 patients from a bond of type p^*h but we do not have such type of bond in $\mathcal{C}$ which contains at least 3 patients to transfer) $\square$

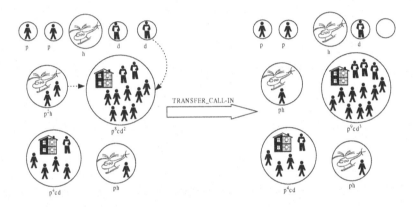

Fig. 1. Collection $\mathcal{C}$ (the left hand side) is modified into collection $\mathcal{C}'$ (the right hand side) after firing the rule TRANSFER_CALL-IN described in Example 2. The dotted arrows in the left hand side of this figure indicate the bonds on which the two atomic rules TRANSFER and CALL-IN (in the rule TRANSFER_CALL-IN) fire upon in parallel.

A bond type L is *regular* if L is a regular language (i.e., a regular language in $a_1^*...a_k^*$). In particular, when L is (regular expression) a, we say that L is the a-free bond type. The object (a) in the a-free bond is called a *free* object.

A *bond computing system (BCS)* M is a machine specified by the following tuple:

$$\langle \Sigma, \mathcal{L}, \mathcal{R}, \theta \rangle, \tag{2}$$

where $\mathcal{L}$ is a finite set of bond types, $\mathcal{R}$ is a finite set of rules, and θ is the *initial constraint*. An *initial* collection of M is one in which a nonempty bond must be of a-free bond type for some a. That is, every object is a free object in an initial collection. Additionally, the initial collection must satisfy the initial constraint θ which is specified by a mapping $\theta : \Sigma \rightarrow \mathrm{N} \cup \{*\}$. That is, for each $a \in \Sigma$ with $\theta(a) \neq *$, the initial collection must contain exactly $\theta(a)$ number of free a-objects (while for a with $\theta(a) = *$, the number of free a-objects in the initial collection could be any number). Hence, the initial constraint θ is to specify the symbols in Σ whose initial multiplicities are fixed. For two collections $\mathcal{C}$ and $\mathcal{C}'$, we say that $\mathcal{C}$ can *reach* $\mathcal{C}'$, written

$$\mathcal{C} \leadsto_M \mathcal{C}'$$

if, for some n,

$$\mathcal{C}_0 \overset{R_1}{\to} \mathcal{C}_1 \cdots \overset{R_n}{\to} \mathcal{C}_n$$

where $\mathcal{C} = \mathcal{C}_0, \mathcal{C}_1, \cdots, \mathcal{C}_n = \mathcal{C}'$ are collections, and $R_1, \cdots, R_n$ are rules in $\mathcal{R}$.

In this paper, we focus on *regular* bond computing system (RBCS) M where each $L \in \mathcal{L}$ is regular. As we mentioned earlier, a BCS M can be thought of a design model of a pervasive software system. An important issue in developing such a design, in particular for safety-critical and/or mission critical systems, is to make sure that the software system does not have some undesired behaviors. This issue is called *(automatic) verification*, which seeks an (automatic) way to verify that a system satisfies a given property.

Example 3. Let M be a regular bond computing system with $\Sigma = \{p, h, c, d\}$ and $\mathcal{L}$ specified in Example 1. The initial constraint is defined as follows: $\theta(p) = *$, $\theta(h) = 3$, $\theta(c) = 2$, and $\theta(d) = 5$. The rules in $\mathcal{R}$ are in below:

- R_1: $(p)_1 \overset{p}{\to} (p^*h)_2$;
- R_2 consisting of two atomic rules:
    r_1: $(p^*h)_1 \overset{p}{\to} (p^8cd^2)_2$;
    r_2: $(d)_3 \overset{d}{\to} (p^8cd^2)_2$;
- R_3: $(p^*h)_1 \overset{p}{\to} (p^{\leq 8}c(d + dd))_2$;
- R_4: $(p^*h)_1 \overset{p}{\to} (p^{>8}cd^{>2})_2$;
- R_5 consisting of two atomic rules:
    r_1: $(p^8cd^3)_1 \overset{p}{\to} (\lambda)_2$;
    r_2: $(p^8cd^3)_1 \overset{d}{\to} (\lambda)_3$;
- R_6: $(p^{\leq 8}c(d + dd))_1 \overset{p}{\to} (\lambda)_2$;
- R_7: $(p^{>8}cd^{>2})_1 \overset{p}{\to} (\lambda)_2$;
- R_8: $(d)_1 \overset{d}{\to} (c)_2$;
- R_9: $(d)_1 \overset{d}{\to} (cd)_2$.

In above, R_1 indicates that a patient boards a helicopter. R_2, R_3 and R_4 indicate that a patient is transferred from a helicopter to a clinic, where, in particular, R_2 means that the clinic does not have enough doctors, so an additional doctor is called in. R_5, R_6 and R_7 indicate that a patient leaves the clinic (after being cured). In particular, R_5 means that, when the clinic does not have many patients, a doctor can be off-duty. R_8 and R_9 are to fill a clinic (without patient) with one or two doctors. □

A simplest (and, probably, the most important) class of verification queries concern *reachability*, which has various forms. A *collection-to-collection reachability* problem is to decide whether, given two collections, one can reach the other in a given BCS:

Given: an RBCS M and two collections $\mathcal{C}$ and $\mathcal{C}'$ (where $\mathcal{C}$ is initial),

Question: $\mathcal{C} \leadsto_M \mathcal{C}'$?

Observe that, since the total number of objects in M does not change in any run of M from the initial collection $\mathcal{C}$, the collection-to-collection reachability problem is clearly decidable in time polynomial in the size of M and the size of $\mathcal{C}$ (the total number of objects in $\mathcal{C}$, where the number is *unary*). Just use "breadth-first" to find all reachable collections.

Theorem 1. *The collection-to-collection reachability problem for regular bond computing systems is decidable in polynomial time.*

One can also generalize Theorem 1 to the *generalized collection-to-collection reachability* problem, where the initial collection is still given but the target collection is not:

Given: an RBCS M, a regular bound type L, and an initial collection $\mathcal{C}$,

Question: is there a collection $\mathcal{C}'$ such that $\mathcal{C} \leadsto_M \mathcal{C}'$ and $\mathcal{C}'$ contains a bond of type L ?

Theorem 2. *The generalized collection-to-collection reachability problem for regular bond computing systems is decidable in polynomial time.*

Remark 1. It is easily verified that the two theorems above hold even when the bond types are not regular, but are in PTIME; i.e., the class of languages accepted by deterministic Turing machines in polynomial time.

In the collection-to-collection reachability problems studied in Theorems 1 and 2, M starts with a given initial collection. Therefore, the decision algorithms for the problems essentially concern testing: each time, the decision algorithms only verify M under a concrete initial collection. It is desirable to verify that the system M has the undesired behavior for *some* initial collections. That is, considering the negation of the problem, we would like to know whether the design is correct under every possible (implementation) instance.

To this end, we will study the following *bond-type reachability* problem, where the initial collection is not given:

Given: an RBCS M and a regular bound type L,

Question: are there two collections $\mathcal{C}$ and $\mathcal{C}'$ such that $\mathcal{C}$ is initial, $\mathcal{C} \leadsto_M \mathcal{C}'$, and $\mathcal{C}'$ contains a bond of type L ?

The answer of the above question is meaningful in many aspects. One is that it can help decide whether a system is well designed. Suppose the bond type L is something undesired which we do not want to have in a system. If we have $\mathcal{C} \leadsto_M \mathcal{C}'$, and $\mathcal{C}'$ contains a nonempty bond of type L, we say that the system is not well designed.

Example 4. Consider the bond computing system M in example 3 and with $\mathcal{C} = \{p, p, h, d, d, p^2h, ph, p^4cd, p^8cd^2\}$ reached at some moment. Suppose that $L = p^{>8}cd^{\leq 2}$ is an "undesired" bond type (e.g., a clinic with too many patients but with too few doctors). First, we choose $\langle w_1, w_2 \rangle = \langle p, p^2h \rangle$ from $\mathcal{C}$. After applying R_1, $\langle w_1', w_2' \rangle = \langle \lambda, p^3h \rangle$ and we get $\mathcal{C}' = \{p, h, d, d, p^3h, ph, p^4cd, p^8cd^2\}$, or, $\mathcal{C} \overset{R_1}{\rightarrow} \mathcal{C}'$. Next, we pick $\langle w_1, w_2 \rangle = \langle p^3h, p^8cd^2 \rangle$ from $\mathcal{C}'$, after applying R_3, $\langle w_1', w_2' \rangle = \langle p^2h, p^9cd^2 \rangle$, we get $\mathcal{C}'' = \{p, h, d, d, p^2h, ph, p^4cd, p^9cd^2\}$, or, $\mathcal{C}' \overset{R_3}{\rightarrow} \mathcal{C}''$. Therefore, $\mathcal{C} \rightsquigarrow_M \mathcal{C}''$, and in $\mathcal{C}''$ there is a bond p^9cd^2, which is a bond of the undesired bond type L. Therefore, M is not designed as expected.   □

In the following sections, we are going to investigate the computing power of regular bond computing systems. Our results show that the bond-type reachability problem is undecidable in general. However, there are interesting special cases where the problem becomes decidable.

3   Undecidability of Bond-Type Reachability

To show that the bond-type reachability problem is undecidable, we first establish that the computing power of RBCS (regular bond computing systems) lies between LBA (linearly bounded automata) and LBC (linearly bounded multi-counter machines). To proceed further, some definitions are needed.

An LBA A is a Turing machine where the R/W head never moves beyond the original input region. For the purpose of this paper, we require that the LBA works on bounded input; i.e., input in the form of $b_1^* \cdots b_m^*$. As usual, we use Lang(A) to denote the set of input words that are accepted by A.

Let M be an RBCS. Recall that, in M, the initial constraint θ is used to specify the case when some symbols in Σ whose initial multiplicities are fixed (and specified by θ). Without loss of generality, we use $a_1, \cdots, a_l$ (for some $l \leq k$) to denote the remaining symbols and, by ignoring those "fixed symbols", we simply use $a_1^{i_1} \cdots a_l^{i_l}$ to denote an initial collection $\mathcal{C}$ of M where each object is a free object. Let L be a given regular bond type. We say that the initial collection is L-*accepted* by M if there is some collection $\mathcal{C}'$ that contains a bond of type L such that $\mathcal{C} \rightsquigarrow_M \mathcal{C}'$. As usual, we use Lang($M, L$) to denote the set of all words $a_1^{i_1} \cdots a_l^{i_l}$ that are L-accepted by M. Notice that the bond type reachability problem is equivalent to the emptiness problem for Lang(M, L).

We first observe that, since the total "size" (i.e., the total number of objects in the system) of a collection always equals the size of the initial collection during a run of M, an RBCS M can be simulated by an LBA.

Theorem 3. *For any given regular bond computing system M and a regular bond type L, one can construct a linearly bounded automaton A such that* Lang(M, L) = Lang(A).

Remark 2. Clearly, the theorem above holds, even when the bond types are not regular, but are context-sensitive languages (= languages accepted by LBAs).

Remark 3. Actually, LBAs are strictly stronger than RBCS (since languages accepted by RBCS are upper-closed (i.e., if $a_1^{i_1} \cdots a_k^{i_k}$ is accepted then so is $a_1^{i'_1} \cdots a_k^{i'_k}$, for any $i_j \leq i'_j$) and this is not true for LBAs).

A multicounter machine is a (nondeterministic) program P equipped with a number of nonnegative integer counters $X_1, \cdots, X_n$ (for some n). Each counter can be incremented by one, decremented by one, and tested for 0, along with state transitions. That is, P consists of a finite number of instructions, each of which is in one of the following three forms:

$$s : X + +, \texttt{goto } s';$$
$$s : X - -, \texttt{goto } s';$$
$$s : X == 0?, \texttt{goto } s';$$

where X is a counter. Initially, the counters start with 0. During a run of the machine P, it crashes when a counter falls below 0. Notice that, in our instruction set, there is no instruction in the form of $s : X > 0?, \texttt{goto } s'$. In fact, this instruction can be simulated by the following two instructions: $s : X--, \texttt{goto } s''$ and $s'' : X++, \texttt{goto } s'$, where s'' is a new state. The machine halts when it enters a designated accepting state. It is well known that multicounter machines are Turing-complete (even when there are two counters). We say that a nonnegative integer tuple $(x_1, \cdots, x_n)$ is *accepted* by P if P has an accepting run during which the maximal value of counter X_i is bounded by x_i for each i. One can treat the tuple $(x_1, \cdots, x_n)$ as an (input) word (where each number is unary). In this way, we use language $\text{Lang}(P)$ to denote the set of all nonnegative integer tuples accepted by P. Notice that, in here, counter values in the multicounter machine can not exceed the values given in the input. Therefore, we call such P as a *linearly bounded multicounter machine* (LBC).

Our next result shows that an LBC P can be simulated by an RBCS M. A rule that contains only one atomic rule is called *simple*. We call M *simple* if every rule in M is simple. In a collection $\mathcal{C}$, a bond is *trivial* if it is either empty or of a free bond type; that is, the bond contains at most one (free) object. When there are at most m nontrivial bonds in any reachable collection during any run of M (starting from any initial collection), we simply say that M *contains m nontrivial bonds*. In fact, the result holds even when M is simple and contains 2 nontrivial bonds.

Theorem 4. *For any given linearly bounded multicounter machine P, one can construct an RBCS M, which contains at most 2 nontrivial bonds, and a regular bond type L such that $\text{Lang}(P) = \text{Lang}(M, L)$.*

An atomic rule in (1) is *single* if there is exactly one object that will be transferred; i.e., the size of α_r is 1. A rule that contains only single atomic rules is also called *single*. An RBCS M is *single* if it contains only single rules. Notice that every simple rule can be rewritten into a single rule. For instance, a simple rule like $(a^+ b^*)_1 \xrightarrow{abb} (a^* c^+)_2$ is equivalent to the single rule consisting of the following three atomic single rules:

$(a^+b^*)_1 \xrightarrow{a} (a^*c^+)_2$;
$(a^+b^*)_1 \xrightarrow{b} (a^*c^+)_2$;
$(a^+b^*)_1 \xrightarrow{b} (a^*c^+)_2$.

Therefore, a simple RBCS can be simulated by a single RBCS. Following the same idea, one can also observe that an RBCS can be simulated by a single RBCS. Hence, combining Theorems 3 (and Remark 3) and Theorem 4, we have the following diagram summarizing the computing power, in terms of language acceptance, of regular bond computing systems (from higher to lower):

$$\text{LBA} > \text{RBCS} = \text{single RBCS} \geq \text{simple RBCS} \geq \text{LBC}.$$

We currently do not know whether some of the $\geq$'s are strict.

Notice that the emptiness problem for LBC (which is equivalent to the halting problem for multicounter machines) is known undecidable. According to Theorem 4, for (single) RBCS, the emptiness problem (i.e., $\text{Lang}(M, L) = \emptyset$? for a given RBCS M and regular bond type L) is undecidable as well. Hence,

Theorem 5. *The bond-type reachability problem for regular bond computing systems is undecidable. The undecidability remains even when the systems are simple.*

We now generalize the notion of L-acceptance for an RBCS M. Let $L_1, \cdots, L_n$, $Q_1, \cdots, Q_m$ (for some n and m) be bond types.

Similar to L-acceptance defined earlier, we say that an initial collection $\mathcal{C}$ is $(L_1, \cdots, L_n; Q_1, \cdots, Q_m)$-*accepted* by M if there is some collection $\mathcal{C}'$ with $\mathcal{C} \leadsto_M \mathcal{C}'$ such that $\mathcal{C}'$ contains a bond of type L_i for each i and does not contain a bond of type Q_j for each j. As in the definition of $\text{Lang}(M, L)$, we use $\text{Lang}(M, L_1, \cdots, L_n; Q_1, \cdots, Q_m)$ to denote the set of all words $a_1^{i_1} \cdots a_l^{i_l}$ that are $(L_1, \cdots, L_n; Q_1, \cdots, Q_m)$-accepted by M.

Now, Theorem 4 can be generalized to show that an RBCS can simulate a *multihead two-way nondeterministic finite automaton* (multihead 2NFA).

A multihead 2NFA M is a nondeterministic finite automaton with k two-way read-only input heads (for some k). An input string $x \in a_1^* \dots a_n^*$ (the a_i's are distinct symbols) is accepted by M if, when given $\$x\$$ (the $\$$'s are the left and right end markers) with M in the initial state and all k heads on the left end marker, M eventually halts in an accepting state with all heads on the right end marker. The language accepted by M is denoted by $\text{Lang}(M)$.

Now let P be an LBC with a set C of k counters. Let $n \geq 1$. Partition C into n sets $C_1, \dots, C_n$ of counters, where each C_i has k_i counters, and $k = k_1 + \dots + k_n$. We say that a nonnegative integer tuple $(x_1, \dots, x_n)$ is accepted by P if P has an accepting run during which each of the k_i counters in C_i is bounded by (**and achieve**) the value x_i. Define $\text{Lang}(P)$ to be the set of all n-tuples $(x_1, \dots, x_n)$ accepted by P. Note that the definition given earlier can be considered as the current definition with $k_1 = \dots = k_n = 1$ but without requiring the bound x_i being achieved.

Slightly modifying the proof of Theorem 4 (using $Q_1, \cdots, Q_m$ to specify that free bond types all disappear at the end of computation), we have

Theorem 6. *For any given linearly bounded multicounter machine P, one can construct an RBCS M, which contains at most 2 nontrivial bonds, and regular bond types $L_1, \cdots, L_n, Q_1, \cdots, Q_m$ such that $\mathrm{Lang}(P) = \mathrm{Lang}(M, L_1, \cdots, L_n, Q_1, \cdots, Q_m)$.*

Actually, LBC and multihead 2NFA are equivalent in the following sense:

Lemma 1. *A set $O \subseteq N^n$ is accepted by an LBC P if and only if the language $\{a_1^{x_1} a_2^{x_2} ... a_n^{x_n} \mid (x_1, ..., x_k) \in O\}$ is accepted by a multihead 2NFA.*

From Theorem 6, we have the following corollary:

Corollary 1. *For any given multhead 2NFA M, one can construct an RBCS M', which contains at most 2 nontrivial bonds, and regular bond types $L_1, \cdots, L_n, Q_1, \cdots, Q_m$ such that $\mathrm{Lang}(M) = \mathrm{Lang}(M', L_1, \cdots, L_n, Q_1, \cdots, Q_m)$.*

Since multihead 2NFAs are equivalent to $\log n$ space-bounded nondeterministic Turing machines (NTMs), we also obtain the following result:

Corollary 2. *For any given $\log n$ space-bounded NTM M, one can construct an RBCS M', which contains at most 2 nontrivial bonds, and regular bond types $L_1, \cdots, L_n, Q_1, \cdots, Q_m$ such that $\mathrm{Lang}(M) = \mathrm{Lang}(M', L_1, \cdots, L_n, Q_1, \cdots, Q_m)$.*

Next, we show that an RBCS can simulate the computation of a nondeterministic linear-bounded automaton (LBA). Let M be an LBA with input alphabet $\Sigma = \{1, 2, ..., k\}$. We represent a string $x = d_n...d_0$ (each d_i in $\{1, ..., k\}$) as an integer $N(x) = d_n k^n + d_{n-1} k^{n-1} + ... + d_1 k^1 + d_0 k^0$. Let o be a symbol. For a language $L \subseteq \Sigma^*$, define $\mathrm{Unary}(L) = \{o^{N(x)} \mid x \in L\}$. In [11], it was shown that given an LBA M, one can construct a multihead 2NFA M' such $\mathrm{Lang}(M') = \mathrm{Unary}(\mathrm{Lang}(M))$. Hence, we have:

Corollary 3. *For any given LBA M, one can construct an RBCS M', which contains at most 2 nontrivial bonds, and regular bond types $L_1, \cdots, L_n, Q_1, \cdots, Q_m$ such that $\mathrm{Lang}(M) = \mathrm{Lang}(M', L_1, \cdots, L_n, Q_1, \cdots, Q_m)$.*

4   Restricted Bond Computing Systems

According to Theorem 5, the bond-type reachability problem is undecidable for regular bond computing systems. In this section, we are going to investigate further restrictions that can be applied such that the problem becomes decidable.

An n-dimensional *vector addition system with states* (VASS) G is a 5-tuple

$$\langle V, p_0, p_f, S, \delta \rangle$$

where V is a finite set of *addition vectors* in Z^n, S is a finite set of *states*, $\delta \subseteq S \times S \times V$ is the *transition relation*, and $p_0, p_f \in S$ are the *initial state* and the *final state*, respectively. Elements (p, q, v) of δ are called *transitions* and are usually written as $p \rightarrow (q, v)$. A *configuration* of a VASS is a pair (p, u) where $p \in S$ and $u \in N^n$. The transition $p \rightarrow (q, v)$ can be applied to the configuration (p, u) and yields the configuration $(q, u + v)$, provided that $u + v \geq \mathbf{0}$ (in this case, we write $(p, u) \rightarrow (q, u + v)$). For vectors x and y in N^n, we say that x can *reach* y, written $x \leadsto_G y$, if for some j,

$$(p_0, x) \rightarrow (p_1, x + v_1) \rightarrow \cdots \rightarrow (p_j, x + v_1 + ... + v_j)$$

where p_0 is the initial state, p_j is the final state, $y = x + v_1 + ... + v_j$, and each $v_i \in V$. It is well-known that Petri nets and VASS are equivalent.

Let B be a constant and M be an RBCS. We say that an RBCS M is B-bounded if, in any reachable collection of M (starting from any initial collection), there is no bond containing more than B objects. Essentially, we restrict the size of each bond in M while we do not restrict the number of the nonempty bonds. Notice that there are at most t (depending only on B and k – the size of the alphabet Σ) distinct words (i.e., bonds)

$$w_0, w_1, \cdots, w_t,$$

where each w_i is with size at most B and w_0 is the empty word. Accordingly, we use L_{w_i} to denote the bond type $\{w_i\}$. For a collection $\mathcal{C}$ of the B-bounded M, we use $\#(w_i)$ to denote the multiplicity of type L_{w_i} (nonempty) bonds in $\mathcal{C}$, with $i \geq 1$. In this way, the $\mathcal{C}$ can then be uniquely represented by a vector $\#(\mathcal{C})$ defined as

$$(\#(w_1), \cdots, \#(w_t)) \in N^t.$$

Accordingly, the reachability relation $\mathcal{C} \leadsto_M \mathcal{C}'$ for the B-bounded M can then be represented by $\#(\mathcal{C}) \leadsto_M \#(\mathcal{C}')$; i.e., $\leadsto_M \subseteq N^t \times N^t$.

Our definition of B-boundedness for RBCS M is effective. That is, for any B-bounded RBCS M, one can construct an equivalent B-bounded RBCS M' such that the nonempty bond types in M' are $L_{w_1}, \cdots, L_{w_t}$ and $\leadsto_M = \leadsto_{M'}$. The proof is straightforward. So, without loss of generality, we further assume that the bond types in a B-bounded RBCS M are given as $L_{w_1}, \cdots, L_{w_t}$.

Theorem 7. *B-bounded regular bond computing systems can be simulated by vector addition systems with states. That is, for a B-bounded RBCS M, one can construct a VASS G such that $\leadsto_M = \leadsto_G$.*

A subset S of N^n is a *linear set* if there exist vectors $v_0, v_1, ..., v_t$ in N^n such that $S = \{v \mid v = v_0 + a_1 v_1 + \cdots + a_t v_t, \forall 1 \leq i \leq t, a_i \in N\}$. The vectors v_0 (the constant vector) and $v_1, ..., v_t$ (the periods) are called *generators*. A set S is *semilinear* if it is a finite union of linear sets. The *semilinear reachability* problem for VASS is as follows: Given a VASS G and semilinear sets P and Q, are there vectors $u \in P$ and $v \in Q$ such that $u \leadsto_G v$? The problem can be easily shown decidable, by making a reduction to the classic (vector to vector) reachability problem (which is known decidable) for VASS.

Lemma 2. *The semilinear reachability problem for VASS is decidable.*

Using Lemma 2 and Theorem 7, it is straightforward to show

Corollary 4. *The bond-type reachability problem for B-bounded RBCS is decidable.*

Notice that, if we remove the condition of B-boundedness from Corollary 4, the bond-type reachability problem is undecidable, as shown in Theorem 5. The undecidability remains even when simple RBCS are considered. An interesting case now is whether the problem becomes decidable when an RBCS M is *both* single and simple. Or, for a more general case, each rule in M contains only one atomic rule and transfers exactly one type of objects(i.e., l a-objects for some l); we call such M as a *1-transfer* RBCS.

The following theorem shows that the regular bond-type reachability problem is decidable for 1-transfer RBCS. Notice that the problem is undecidable if the 1-transfer condition is removed, as shown in the construction of Theorem 4. The key idea used in the following proof is to show that under a proper symbolic representation of collections, the backward reachability calculation converges.

Theorem 8. *The regular bond-type reachability problem for 1-transfer RBCS is decidable.*

5   Conclusion

In this paper, we introduced Bond Computing Systems (BCS) to model high-level dynamics of pervasive computing systems. The model is a variation of P systems introduced by Gheorghe Paun and, hence, is biologically inspired. We focused on regular bond computing systems (RBCS), where bond types are regular, and study their computation power and verification problems. Among other results, we showed that the computing power of RBCS lies between LBA (linearly bounded automata) and LBC (a form of bounded multicounter machines) and hence, the bond-type reachability problem (given an RBCS, whether there is some initial collection that can reach some collection containing a bond of a given type) is undecidable. We also study some restricted models (i.e., B-boundedness and 1-transfer) of RBCS where the reachability problem becomes decidable.

Notice that our model of BCS is not universal. Clearly, if one generalizes the model by allowing rules that can create an object from none, then the model becomes Turing-complete (from the proof of Theorem 4). In this paper, we only focus on regular BCS. In the future, we will further investigate "nonregular" BCS where bond types involve linear constraints. We are currently implementing a design language based on BCS and study an automatic procedure that synthesizing a pervasive application instance running over a specific network from a BCS specification.

References

1. Cook, D.: Health monitoring and assistance to support aging in place. Journal of Universal Computer Science 12(1), 15–29 (2006)
2. Dang, Z., Ibarra, O.H.: On P systems operating in sequential mode. In: Preproceedings of the 6th Workshop on Descriptional Complexity of Formal Systems (DCFS'04), 2004. Report No. 619, Univ. of Western Ontario, London, Canada, pp. 164–177 (2004)
3. Ibarra, O.H., Yen, H., Dang, Z.: The power of maximal parallelism in p systems. In: Calude, C.S., Calude, E., Dinneen, M.J. (eds.) DLT 2004. LNCS, vol. 3340, pp. 212–224. Springer, Heidelberg (2004)
4. Kumar, M., Shirazi, B., Das, S.K., Sung, B.Y., Levine, D., Singhal, M.: PICO: a middleware framework for pervasive computing. Pervasive Computing, IEEE 2(3), 72–79 (2003)
5. Lamport, L., Lynch, N.: Distributed computing: models and methods. Handbook of theoretical computer science (vol. B): formal models and semantics, 1157–1199 (1991)
6. Martin-Vide, C., Paun, Gh., Pazos, J., Rodriguez-Paton, A.: Tissue P systems. Theoretical Computer Science 296(2), 295–326 (2003)
7. Park, S.H., Won, S.H., Lee, J.B., Kim, S.W.: Smart home-digitally engineered domestic life. Personal and Ubiquitous Computing 7(3-4), 189–196 (2003)
8. Paun, Gh.: Computing with membranes. Journal of Computer and System Sciences 61(1), 108–143 (2000)
9. Paun, Gh.: Introduction to membrane computing. See P Systems Web Page at (2004), http://psystems.disco.unimib.it
10. Roman, G.-C., Julien, C., Payton, J.: A formal treatment of context-awareness. In: Wermelinger, M., Margaria-Steffen, T. (eds.) FASE 2004. LNCS, vol. 2984, pp. 12–36. Springer, Heidelberg (2004)
11. Savitch, W.J.: A note on multihead automata and context-sensitive languages. Acta Informatica 2(3), 249–252 (1973)
12. Schilit, B., Adams, N., Want, R.: Context-aware computing applications. In: Proceedings of Workshop on Mobile Computing Systems and Applications, pp. 85–90. IEEE Computer Society Press, Santa Cruz, CA (1994)
13. Sung, B.Y., Kumar, M., Shirazi, B., Kalasapur, S.: A formal framework for community computing. Technical report (2003)
14. Weiser, M.: The computer for the 21st century. Scientific American 265(3), 66–75 (1991)
15. Zimmer, P.: A calculus for context-awareness. BRICS Research Series (2005)

Author Index

Lecture Notes in Computer Science

For information about Vols. 1–4542

please contact your bookseller or Springer